NUTRIENTS IN NATURAL WATERS
    Herbert E. Allen and James R. Kramer, Editors

pH AND pION CONTROL IN PROCESS AND WASTE STREAMS
    F. G. Shinskey

INTRODUCTION TO INSECT PEST MANAGEMENT
    Robert L. Metcalf and William H. Luckman, Editors

OUR ACOUSTIC ENVIRONMENT
    Frederick A. White

ENVIRONMENTAL DATA HANDLING
    George B. Heaslip

THE MEASUREMENT OF AIRBORNE PARTICLES
    Richard D. Cadle

ANALYSIS OF AIR POLLUTANTS
    Peter O. Warner

ENVIRONMENTAL INDICES
    Herbert Inhaber

URBAN COSTS OF CLIMATE MODIFICATION
    Terry A. Ferrar, Editor

CHEMICAL CONTROL OF INSECT BEHAVIOR:
    THEORY AND APPLICATION
    H. H. Shorey and John J. McKelvey, Jr.

MERCURY CONTAMINATION: A HUMAN TRAGEDY
    Patricia A. D'Itri and Frank M. D'Itri

POLLUTANTS AND HIGH RISK GROUPS
    Edward J. Calabrese

Pollutants and High-Risk Groups

About the author

EDWARD J. CALABRESE is Assistant Professor of Environmental Health at the University of Massachusetts. From 1974 to 1976 he was Assistant Professor of Occupational and Environmental Medicine at the University of Illinois, and served as Assistant Director of the State of Illinois' Environmental Health Resource Center. He is the author of numerous papers in his field.

Pollutants and High-Risk Groups

The Biological Basis of
Increased Human Susceptibility
to Environmental
and Occupational Pollutants

Edward J. Calabrese
University of Massachusetts
Amherst, Massachusetts

A Wiley–Interscience Publication
JOHN WILEY AND SONS
New York   Chichester   Brisbane   Toronto

Library of Congress Cataloging in Publication Data

Calabrese, Edward J   1946–
  Pollutants and high risk groups.

  (Environmental science and technology)
  "A Wiley-Interscience publication."
  Bibliography: p. 206
  Includes index.
  1. Pollution—Toxicology.   2. Environmentally
induced diseases—Etiology.   I. Title.

RA566.C29   614.7   77–13957
ISBN 0–471–02940–8

Printed in the United States of America

10 9 8 7 6 5 4 3 2 1

To Mary,
my parents
and the memory of
Robert MacNamara

Series Preface

Environmental Science and Technology

The Environmental Science and Technology Series of Monographs, Textbooks, and Advances is devoted to the study of the quality of the environment and to the technology of its conservation. Environmental science therefore relates to the chemical, physical, and biological changes in the environment through contamination or modification, to the physical nature and biological behavior of air, water, soil, food, and waste as they are affected by man's agricultural, industrial, and social activities, and to the application of science and technology to the control and improvement of environmental quality.

The deterioration of environmental quality, which began when man first collected into villages and utilized fire, has existed as a serious problem under the ever-increasing impacts of exponentially increasing population and of industrializing society. Environmental contamination of air, water, soil, and food has become a threat to the continued existence of many plant and animal communities of the ecosystem and may ultimately threaten the very survival of the human race.

It seems clear that if we are to preserve for future generations some semblance of the biological order of the past and hope to improve on the deteriorating standards of urban public health environmental science and technology must quickly come to play a dominant role in designing our social and industrial structure for tomorrow. Scientifically rigorous criteria of environmental quality must be developed. Based in part on these criteria, realistic standards must be established and our technological progress must be tailored to meet them. It is obvious that civilization will continue to require increasing amounts of fuel, transportation, industrial chemicals, fertilizers, pesticides, and countless other products and that it will continue to produce waste products of all descriptions. What is urgently needed is a total systems approach to modern civilization through which the pooled talents of scientists and engineers, in cooperation with social scientists and the medical profession, can be focused on the development of order and equilibrium to the presently disparate segments of the human environment. Most of the skills and tools that are needed are already in existence. Surely a technology that has created such manifold environmental problems is also capable of solving them. It is our hope that this series in environmental sciences and technology will not only serve to make this challenge more explicit to the established professional but that it also will help to stimulate the student toward the career opportunities in this vital area.

Robert L. Metcalf
James N. Pitts, Jr.
Werner Stumm

Preface

The intention of this book is to make the general public, scientists, and governmental decision makers aware of the effects of pollutants on the hypersusceptible segments of the population so that more informed decisions can be made in the areas of environmental and occupational health, especially with regard to standard setting. This book represents the first comprehensive synthesis of the biomedical literature in which the numerous hypersusceptible segments of the population with respect to pollutant toxicity have been both identified and quantified. In addition to reviewing previously recognized high-risk groups, numerous new high-risks groups have been identified, which may have important implications for future environmental and occupational health policy. Biological factors such as certain developmental processes (Chapter 2), genetic disorders (Chapter 3), nutritional deficiencies (Chapter 4), disease processes (Chapter 5), and behavioral factors (Chapter 6) which predispose individuals to the toxic effects of pollutants are discussed in detail. Chapter 8 reviews the likelihood of medical surveillance for high-risk groups, especially with regard to the industrial setting. Finally, Chapters 8 and 9 interpret the role of high-risk groups in the development of environmental and occupational health policy with regard to standard setting (environmental and occupational; carcinogens and noncarcinogens), economic health cost assessment (with regard to both standard setting and new technology), environmental impact statements, and the new toxic substances control act.

I would like to sincerely thank Dr. Bertram W. Carnow, Professor and Chairman of Occupational and Environmental Medicine and Director of the Environmental Health Resource Center at the University of Illinois, School of Public Health, for his encouragement, support, and guidance with respect to my professional and personal growth. It was Dr. Carnow who first introduced me to the concept of high-risk groups and their importance in the evaluation of the effects of pollutants on human health. I would also like to acknowledge the kind support of the entire staff of the Environmental Health Resource Center, especially Dr. Badi Boulos, Mr. Alfred Sorensen, Mr. William Kojola, and Ms. Sally Jansen.

Acknowledgments are also in order for the faculty of the Environmental Health Program and Epidemiology/Biostatistics Program at the Division of Public Health, University of Massachusetts. Particular gratitude is directed toward Dr. Salvatore DiNardi and Dr. Robert W. Tuthill.

Finally, I would like to thank my wife, Mary, for her encouragement throughout this endeavor.

EDWARD J. CALABRESE

Amherst, Massachusetts
January 1978

Contents

Figures

Tables

Pollutants and
High-Risk Groups

Introduction

1

Environmental health standards have been designed to protect human as well as plant and animal life. The question naturally arises as to what percentage of the population is actually being protected from the toxic or carcinogenic activities of the pollutant.

The Clean Air Act of 1970 demands that the primary air quality standards be such as to ensure full protection to both specifically susceptible subgroups and healthy individuals of the population (Finklea et al., 1974). A number of efforts have been made to identify some of the high-risk* segments of the population, especially with regard to genetic (Stokinger and Scheel, 1973; NAS, 1975) and nutritional deficiencies (Shakman, 1974). However, at present, the identification and especially the quantification of the numbers of individuals at high risk in the population in question is still in its rudimentary stages.

Despite the legislative intention to protect even those who are highly susceptible to the adverse effects of pol-

---

* To be at high risk with respect to a pollutant, an individual would experience the adverse health effects of the pollutant significantly before the general population because of some genetic, developmental, nutritional, physiological, behavioral, psychological, or disease state factors present which predispose the individual to the harmful effects (Table 1).

lutants, practical realities have usually taken precedence. Thus, ultimately, a cost-benefit analysis is attempted in the process of standard derivation in the United States. It is generally easy for industry to determine the cost in dollars of certain required new control devices, but it is very difficult to quantify the cost in human health of proposed standards. Both types of information are critical for intelligent and farsighted decision making. What is critical, yet difficult to provide, is

Table 1. Biological Factors Predisposing Individuals to the Toxic Effects of Environmental-Occupational Pollutants

I. Developmental
   A. Immature enzyme detoxification systems
   B. Immature immune system
   C. Deficiency in immune response due to aging
   D. Other aging factors (e.g., decline in renal function)
   E. Pregnancy
   F. Circadian rhythms

II. Genetic
   A. Red blood cell disorders
      1. Sickle-cell anemia and trait
      2. Glucose-6-phosphate dehydrogenase deficiency
      3. Other deficiencies, glycolytic and HMP pathways
      4. Catalase deficiency
      5. Methemoglobin reductase deficiency
      6. Thalassemias
      7. Porphyrias (also homeostatic-regulatory disorder)
   B. Serum disorders
      1. Cholinesterase variants
      2. Serum a_1-antitrypsin deficiency
   C. Homeostatic-regulatory disorders
      1. Cystinosis
      2. Cystinuria
      3. Tyrosinemia
      4. Wilson's disease
      5. Cystic fibrosis (also malabsorptive disorder)
      6. Crigler-Najjar syndrome
      7. Gilbert's syndrome
      8. Porphyrias
   D. Immunological disorders
      1. Immunoglobin A deficiency
      2. Hypersensitivity to organic chemicals (isocyanates)

Table 1 (Continued)

    E. Malabsorptive disorders
      1. Cystic fibrosis
      2. Acanthocytosis
    F. Other
      1. Aryl hydrocarbon hydroxylase induction
      2. Carbon disulfide sensitivity
      3. Chloroform toxicity
      4. Sulfite oxidase deficiency
      5. Leber's optic atrophy
      6. Albinism
      7. Phenylketonuria

III. Dietary deficiencies
    A. Vitamins—A, B, C, D, E
    B. Minerals—Ca, Fe, Mg, P, Se, Zn
    C. Proteins—amino acids
    D. Fats
    E. Carbohydrates

IV. Diseases
    A. Heart-lung
    B. Kidney
    C. Liver

 V. Behavioral
    A. Smoking
    B. Drinking
    C. Drug habits
    D. Eating habits

---

information concerning the potential health effects on specific percentages of the population of various possible standards for environmental pollutants. Thus, it is the intention of this document to identify as well as quantify those individuals who because of various biological factors may be predisposed to the toxic or carcinogenic effects of specific environmental-occupational pollutants.

Developarental Processes

2

IMMATURE ENZYME DETOXIFICATION SYSTEMS

Liver Metabolism during Gestation

Embryos and fetuses are often exposed to a variety of substances either inadvertently or deliberately administered (via food, water, air, or medicine) to pregnant women. Some of these substances are intended for the developing child and are not of concern here. However, many other substances are intended only for the mother or are unintentional environmental pollutants.

"Placenta selectivity" controls the rates of chemical transfer from the mother to the embryo or fetus for a broad range of chemicals. The rate of entry is thought to be primarily influenced by the fat solubility of the nonionized chemicals. Of secondary importance are the concentration gradients and the size of the molecules. Usually, drugs with a size of less than 600 molecular weight (mw) units often pass freely across the placenta, but substances with a size of greater than 1000 mw usually do not cross the placenta. Substances with a high degree of dissociation penetrate the placenta quite slowly (Hunt, 1975). A listing of some of the substances which pass the placenta and have shown teratogenic effects includes aluminum (Al), benzene, cadmium (Cd), carbaryl, carbon tetrachloride (CCl_4), chromium (Cr) compounds, copper (Cu), 4-dimethylaminoazobenzene, 2,3-dinitrophenol, fluorine (F), formaldehyde, lead (Pb), malathion, mercury (Hg), nickel (Ni), nitrogen oxides (NO_x), paraquat, parathion, polychlorinated biphenyls (PCBs), and selenium (Se) (Hunt, 1975).

Toxicological studies of foreign substances which pass through the placenta have revealed that there are distinct physiological differences between the developing fetus and adults with regard to the capacity of the liver to metabolize foreign substances. The prevailing view about compound metabolism in the human fetal liver is that the metabolizing enzymes are absent or have negligible activities (Rane et al., 1973). This perspective is generally based on data obtained from animal studies and extrapolated to man.

Compounds metabolized during fetal life have been investigated in various species of animals, usually at the very end of gestation. Only negligible metabolic activity has been found in the fetus, but the ability to metabolize foreign compounds increases postnatally, with rates dependent on the species and the substrate (Jondorf, Maickel, and Brodie, 1958; Fouts and Adamson, 1959; Done, 1964; Short and Davies, 1970; and Feuer and Liscio, 1970).

Systematic studies of the development of drug metabolic processes in the human fetus cannot be performed, since legal abortions are usually permitted only during the first 20 weeks of gestation. Scanty observations have been reviewed on fetal compound metabolism during the second half of human gestation. In these observations, tissue specimens from stillborn babies have been used, but the relevance of such findings for the *in vivo* situation may be questioned (Boulos, 1976).

Despite the limited data concerning human fetal liver metabolism, the available research does support the previously cited animal studies; that is, the rate of metabolism of various compounds by the human fetal liver is much slower as compared with the adult. For example, it has been shown by Pelkonen et al. (1973) that the human fetal liver metabolizes 3,4-benzpyrene, aniline, aminopyrine, and hexobarbital at approximately 2.4 to 36.1 percent of adult liver metabolism. Similarly, the research of other investigators (Ackermann, 1971; Pelkonen et al., 1971; Rane and Ackermann, 1971, 1972; Von Bahr and Borga, 1971; Yaffe et al., 1970) has also indicated that the metabolism of different compounds in the human fetal liver was about 35 to 40 percent of that of adult's liver. These results imply that fetuses may be more susceptible to the toxic effects of these compounds than adults. Furthermore, if a compound is a potent carcinogen which crosses the placenta, the fetus's reduced capacity to metabolize such a compound becomes a potentially serious problem. However, if the compound crossing the placenta is a "precarcinogen" and must be converted to the active carcinogenic state by cellular enzymes, the reduced metabolic activity may actually be a distinct advantage.

The drug-metabolizing activity of liver microsomes (these are structures derived from endoplasmic reticulum on liver homogenization) was first discovered by Brodie and his associates in 1958. Most of the compounds in the process of detoxification or bioactivation in the liver are oxidized. It is now fairly well established that the microsomal oxidative system consists of at least two catalytic components: (1) a cytochrome called P-450 and (2) a flavoprotein catalyzing the reduction of this cytochrome by NADPH, termed NADPH–cytochrome P-450 reductase.

This electron transport (oxidative system) can be diagramatically presented as follows (Rane et al., 1973):

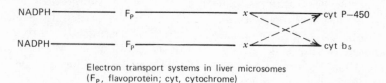

Electron transport systems in liver microsomes
(F_P, flavoprotein; cyt, cytochrome)

Cytochrome P-450 is characterized by a sensitivity to carbon monoxide (CO), which binds to the reduced form of the cytochrome, giving a characteristic absorption spectrum with a maximum of 450 nm (Omura and Sato, 1964, 1964a; Klingenberg, 1958; and Garfinkel, 1958). The flavoprotein NADPH–cytochrome P-450 reductase, in the purified form, has the property of catalyzing a one-electron transfer from NADPH to cytochrome C, which serves as an artificial electron acceptor of the enzyme. It has been shown that in the presence of NADPH and molecular oxygen, liver microsomes catalyze the oxidation of a variety of compounds (Brodie et al., 1958). Depending on the chemical nature of the compound, the reaction may lead to the oxidation of an aromatic ring or a hydrocarbon side chain, an oxidative dealkylation or deamination, or the formation of a sulfoxide. As an example of these processes, Wakabayashi and Shimazono (1961), Robbins (1961), and Das et al. (1968) found the following: liver homogenates supplemented with NADP or NADPH in the presence of oxygen catalyzed the oxidative demethylation of the carcinogenic dye 4-dimethylaminoazobenzene to 4-aminoazobenzene and formaldehyde and the hydroxylation of aliphatic hydrocarbons; the gamma oxidation (γ oxidation) of fatty acids also occurs in such a system. The process may be described by the general reaction

$$RH + NADPH + H^+ + O_2 \rightarrow ROH + NADP + H_2O$$

There is also evidence for an interaction of the NADPH-linked hydroxylating system with NADH–cytochrome C reductase consisting of the flavoprotein NADH–cytochrome b_5 reductase and cytochrome b_5 (Strittmatter, 1965). These above-mentioned oxidizing enzymes are well developed in the adult human liver and the livers of adults of different species. With regard to the human fetal liver, between the seventh and ninth weeks of gestation, the reticulum of the hepatocytes consists of a system of short tubules with ribosome-studded membranes. By approximately the third month of gestation, an increase in the endoplasmic reticulum becomes obvious in the hepatocytes along with deposits of glycogen and iron (Zamboni, 1965). Cytochrome P-450 and flavoprotein NADPH–cytochrome C reductase have also been commonly detected in livers of fetuses in early gestational stages. However, there has been no correlation between levels of enzymes and gestational age.

In general, oxidations can be catalyzed by enzymes in the microsomal fraction of the liver as well as nonmicrosomal enzymes in mitochondria, in soluble fractions of homogenized tissue, and in plasma. Chemically, the major oxidations can be classified as follows (Boulos, 1976):

1. Microsomal oxidations

   a. Aromatic hydrocarbon hydroxylation (to phenols)
   b. Aliphatic hydrocarbon oxidations (to alcohols, ketones, and acids)
   c. *N*-dealkylation (alkyl group oxidation)
   d. *O*- and *S*-dealkylation (alkyl group oxidation)
   e. Epoxidation (to epoxides)
   f. *N*-oxidation (to hydroxylamine or amine oxide derivatives)
   g. *S*-oxidations (to sulfoxides and sulfones)

2. Nonmicrosomal oxidations

   a. Amine oxidation (to aldehydes and ketones)
   b. Alcohol and aldehyde oxidations (to aldehydes and acids)

Table 2 summarizes the known hepatotoxins affecting various

organelles. Most of the reactions involve oxidation through either microsomal or nonmicrosomal systems.

Table 2. .Examples of Hepatotoxins Affecting Various Organelles

Organelles Affected	Compound	Reference
Endoplasmic Reticulum	Carbon tetrachloride	Recknagel, 1967
	Thioacetamide	Thoenes and Bannasch, 1962
	Dimethylnitrosamine	Magee and Swann, 1969
	Phosphorus	Ghoshal et al., 1969
	Ethionine	Anthony et al., 1962
	Dimethylaminoazobenzene	Rouiller, 1964
	Allyl formate	Rouiller, 1964
	Pyrrolizidine alkaloids	McLean, 1970
Mitochondria	Carbon tetrachloride	Recknagel, 1967
	Pyrrolizidine alkaloids	McLean, 1970
	Ethionine	Anthony et al., 1962
	Allyl formate	Rouiller, 1964
	Tannic acid	Rouiller, 1964
	Phosphorus	Rouiller, 1964
	Hydrazine	Ganote and Rosenthal, 1968
	Dimethylnitrosamine	Rouiller, 1964
Lysosomes	Carbon tetrachloride	Dianzani, 1963
	Pyrrolizidine alkaloids	McLean, 1970
	Beryllium	Witschi and Aldridge, 1967
Nucleus	Pyrrolizidine alkaloids	McLean, 1970
	Dimethylnitrosamine	Emmelot and Benedetti, 1960
	Hydrazine	Ganote and Rosenthal, 1968
	Beryllium	Witschi, 1970
	Aflatoxin	Wogan, 1969

Source. Boulos (1976).

Undeveloped Metabolic Pathways versus PCB Toxicity

Since the late 1960s, when PCBs were first recognized as a potentially widespread environmental contaminant, hundreds of research papers have been published concerning their structure, industrial uses, presence in the environment, toxicity to numerous animals from insects to man, tissue storage, metabolism, and excretion (Golberg, 1974;

Selikoff, 1972; Calabrese and Sorensen, 1977). As a result of such data accumulation, regulatory agencies are now trying to approach the problem of standard setting with respect to PCBs. Standards for air, water, and food are necessary if the total human exposure is to be regulated and controlled. One important component in standard setting is a consideration of the individuals in the population who may be at high risk to the pollutant because of various genetic, physiological, psychological, and behavioral traits. Calabrese (1977) has indicated that individuals lacking the ability to detoxify and excrete PCBs represent such a high-risk group, since they are unable to prevent undue accumulation of possible toxic levels of PCBs in the body. The theoretical foundations on which this high-risk group is based (Calabrese, 1977) are subsequently discussed.*

Mammals, including rabbits, dogs, and humans, have been reported to excrete PCBs in part by conjugate glucuronidation or sulfonation (Golberg, 1974; Block and Cornish, 1959). The urinary excretion of biphenyl and 4-chlorobiphenyl has been studied in rabbits. Biphenyl glucosiduronic acid and 4-hydroxybiphenyl were isolated from the urine. The rabbits fed the 4-chlorobiphenyl excreted 4-(p-chlorobiphenyl)-phenol and 4-chlorobiphenyl glucosiduronide. Twice as much 4-chlorobiphenyl as biphenyl was excreted as the glucosiduronic acid derivative. It was suggested that other low-chlorinated biphenyls are excreted in a similar manner.

In dogs injected with 2,4,4'-trichloro-2'hydroxydiphenyl ether, nearly 100 percent of the material excreted in their urine and feces over a 5-day period appeared as the glucuronide or sulfate conjugate. Human adults similarly exposed excrete 65 percent in urine and 20 percent in feces as a free compound or as a glucuronide (Golberg, 1974). Other evidence suggests that some chlorinated biphenyls are hydroxylated by the rat and pigeon. No evidence of reductive dechlorination was observed in either the trout, rat, or pigeon (Hutzinger et al., 1972).

It is apparent that individuals lacking the ability to form either conjugate glucuronides or sulfates or both are at high risk with respect to toxic compounds of a phenolic nature. The recognition of low levels of glucuronide formation in the cat has been suggested as an explanation of the widely observed phenomenon in veterinary literature that phenolic compounds should not be used on or around cats because of the increased incidence of toxicity in this species (Clarke and Clarke, 1967;

* The following section has been previously published as Calabrese, E. J. (1977). Inefficient conjugate glucuronidation: a possible factor in polychlorinated biphenyl (PCB) toxicity. *Medical Hypotheses* 3(4): 162–165; with permission of the publisher of *Medical Hypotheses*.

Jones, 1965; Stecher, 1968; Wilkenson, 1968). Furthermore, intraven-
ous injections of phenol in cats, pigs, dogs, and goats revealed that cats
were at least two times more sensitive with respect to fatality caused by
the phenol (Oehme, 1969). Toxicity in this study was related to the
partial deficiency in the cat to conjugate phenol with glucuronic acid.
The author concluded that the cat is more likely to be poisoned by acute
doses of phenol and is more likely to become chronically affected with
phenol toxicity due to the inability to form and excrete glucuronides
rapidly. Biochemical studies with the cat have also indicated that conjugate glucuronidation is the major pathway for the detoxification and
excretion of phenol (e.g., more than twice as effective as sulfonation)
(Oehme, 1969). Because cats are hypersusceptible to phenoliclike compounds as a result of their diminished capacity to form glucuronides, it
is strongly suspected that human embryos, fetuses, and neonates (2 to 3
months old) which also are deficient in a functional conjugate
glucuronidation system are also predisposed to the toxic effects of
phenolic compounds, including PCBs (Gillette, 1967; Smith and Williams, 1966; Nyhan, 1961).

Usually by the age of 2 to 3 months, adult levels of most enzyme
systems are achieved in humans, as previously suggested (Gillette,
1967; Smith and Williams, 1966; Nyhan, 1961). Unfortunately, this
"developmental immaturity" in the unborn and the very young may
predispose them to the toxic effects of certain substances, since they may
be unable to detoxify and excrete them as quickly as necessary (Nyhan,
1961). It has already been pointed out that human adults do partially
excrete PCBs via conjugate glucuronidation. Clinical experience has
shown that infants may respond differently to doses of drugs that are
easily tolerated by older children and adults. Presumably, this is
because older children and adults have fully functioning enzyme detoxification systems, but the neonate lacks such development.

The widespread occurrence of glucuronic acid conjugation may be due
to the facility with which glucuronic acid can be produced in the body
from carbohydrate sources and the variety of chemical groups to which
glucuronic acid can be transferred enzymatically. The other conjugation
mechanisms are more restricted in their occurrence, probably because
of the limited availability of the conjugating agents such as glycine and
cysteine via glutathione, and the small number of chemical groupings
to which conjugating agents such as sulfate, glycine, cysteine, methyl,
and acetyl can be transferred. The activity and amount of the trans-
ferring enzymes concerned in these conjugations are probably also
limiting factors. Glycine conjugation is confined mainly to aromatic
carboxyl groups; cysteine or glutathione conjugation to some aromatic

hydrocarbons and halogenated hydrocarbons; methylation to certain hydroxyl and amino groups; and acetylation to some amino and hydrazine groups. However, glucuronic acid conjugation can occur with several types of hydroxyl, amino, and carboxyl groups and with sulfhydryl groups (Adamson and Davies, 1973; Dutton, 1961, 1962; Gillette, 1967). Figure 1 illustrates the main types of conjugation (with glucuronic acid, glycine, and sulfate).

The effect of conjugation of a substance with glucuronic acid is to produce a strongly acidic compound which is more water-soluble at physiological pH values than the precursor. The majority of foreign substances is ultimately cleared from the body via their excretion in the urine and bile, most often in the form of polar conjugates such as glucuronides, with the avenue of excretion of the glucuronide probably varying with the species of animal considered (Adamson and Davies, 1973; Dutton, 1962). Figure 2 represents a schematic summary of the successive biochemical modifications which change the physicochemical characteristics of foreign substances, usually resulting in hydrophilicity and thus rapid excretion in the urine.

A tragic example of toxicity arising from the inability to form glucuronides has been reported. The drug chloramphenicol, which is known to be metabolized in humans by the action of the glucuronic pathway, caused the death of more than 30 premature babies who had been treated with the antibiotic for infectious diseases. Theoretically, the premature babies were not able to conjugate the drug with glucuronic acid, and thus the drug could only be slowly excreted and so tended to accumulate in the body to toxic proportions (Smith and Williams, 1966). Inefficient glucuronidative mechanisms in the very young have broad significance, since the absence of such mechanisms prevents the prompt elimination of toxic substances from the body.

A further problem for the very young with regard to PCBs is that in addition to having a difficult time excreting PCBs they may also consume more PCBs per unit of body weight than at any other time during their life span. For example, samples of human milk from two cities in California contained average PCB levels of 60 ppb, but average levels in human milk in Sweden and Germany were 16 and 100 ppb, respectively. Based on a daily milk intake of 150 ml, breast-fed infants in California would ingest about 9μg/kg/day of PCBs. A range of 1 to 3 μg/kg/day has been suggested as a "reasonable" level for an acceptable daily intake (Selikoff, 1972; Berglund, 1967). One μg/kg/day has been reported as being 100 times less than the lowest "no effect" level in animal studies (Vos, 1972). Thus, in children, the margin is reduced to about a factor of 10 if excretory activity is equivalent with adults.

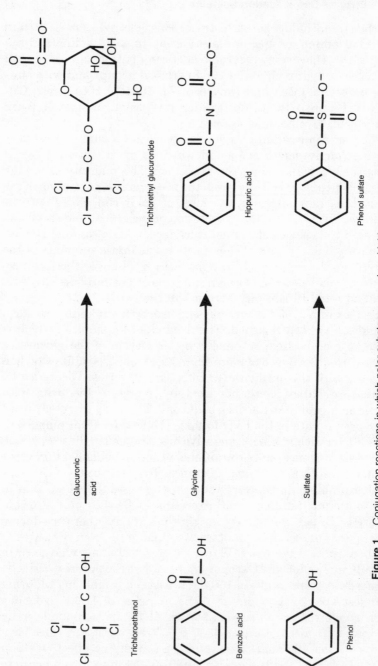

Figure 1. Conjugation reactions in which polar substances produce highly water-soluble metabolites that are rapidly excreted in the kidney, largely by active excretion. The conjugates are strong acids and therefore highly ionized in the urine. As a result, they are not passively reabsorbed. *Source.* Ariens et al. (1976).

Trichlorethyl glucuronide

Hippuric acid

Phenol sulfate

Glucuronic acid

Glycine

Sulfate

Trichloroethanol

Benzoic acid

Phenol

However, as a result of the immature enzyme systems the safety factor of 10 may be considerably reduced.

An additional problem encountered by many neonates is that approximately 5 percent of the mothers of normal infants secrete milk which inhibits the activity of glucuronyl transferase (thus the glucuronidation process) by more than 20 percent via the action of a steroid (pregname 3α, 20β-diol) present in the breast milk. Inhibition of the glucuronyl transferase has been reported for up to 49 days after birth. Clinically, these children have been reported to develop unusually severe neonatal jaundice. This condition develops because glucuronide formation, which assists in the elimination of bilirubin (breakdown product of hemoglobin), is partially inhibited. Cow's milk does not contain sufficient amounts of this steroid to effect a noticeable inhibition of the glucuronidation process. Consequently, about 5 percent of the breast-fed neonates would be expected to have their ability to excrete PCBs further impaired (Garther and Arias, 1966). For a contrary view see Ramos et al. (1966). Administration of the antibiotic novobiocin has also been associated with unconjugated hyperbilirubinemia in infants (Sutherland and Keller, 1961). Novobiocin is a noncompetitive inhibitor of glucuronyl transferase activity *in vitro* (Lokietz et al., 1963). Thus, children receiving concomitant exposure of novobiocin and PCBs would be expected to have their ability to detoxify and excrete PCBs impaired (Calabrese, 1977).

After the neonatal period, a broad range of conditions is associated with the improper or incomplete development of the glucuronide conjugation system. The usual physiological problem associated with these conditions is the inadequate detoxification and excretion of bilirubin. This spectrum extends from the frequently occurring "mild" condition known as Gilbert's syndrome to the very rare, but severe and often fatal, Crigler-Najjar syndrome (Lester and Schmid, 1964).

Since the clinical effects associated with these syndromes are considered to be caused by metabolic disturbances of the glucuronide conjugation scheme, it is expected that PCB elimination in these individuals would be impeded. The population incidence of Gilbert's syndrome has been variously reported as 1 in 200 males (Billing, 1970), 7 percent based on examination of 100 medical students (Kornberg, 1942) and 6 percent of 252 healthy National Blood Transfer Service donors, and 47 healthy medical students (197 males and 102 females) with no difference in frequency between males and females (Owens and Evans, 1975).

Concomitant exposure to PCBs and other druglike chemicals may potentiate the toxic effects of PCBs. For example, rats and monkeys

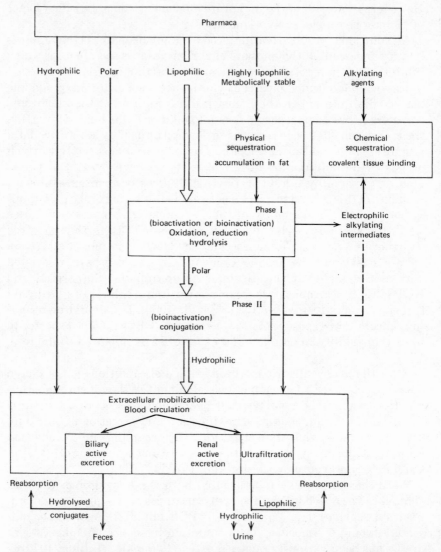

Figure 2. Schematic representation of the most important aspects of pharmacon metabolism. The general trend is a transformation of lipophilic substances to more hydrophilic substances, where toxicity is reduced and excretion via urine is enhanced. *Source.* Ariens et al. (1976).

given SKF525A (β-diethylaminoethyl-2,2-diphenylpentanoate, a nonspecific inhibitor for many of the microsomal enzymes, especially hepatic microsomal enzymes) during the initial 24 hr of exposure to PCB succumbed rapidly as compared with the contr..l group (Allen, 1975). Of possible significance is the fact that SKF525A inhibits the proper functioning of the glucuronic pathway (Smith and Williams, 1966).

Individuals with liver infections may also be at high risk with respect to PCBs. For example, depression of glucuronide synthesis has been observed in humans with infectious hepatitis (Smith and Williams, 1966).

Even though PCB-like pollutants require a functional conjugate glucuronidation system for their excretion, this does not imply that such substances are merely passive molecules in this process. For example, synthesis of microsomal enzymes such as glucuronyl transferase can be stimulated by drugs such as phenobarbital. Phenobarbital treatments have been successfully used with patients with a partial defect in bilirubin conjugation in order to reduce the levels of bilirubin in the blood (Yaffe et al., 1966; Smith et al., 1967; Arias et al., 1969; Kreek and Sleisenger, 1968). Thompson et al. (1969) demonstrated similar results with oral administration of dichlorodiphenyl trichloroethane (pp'DDT). Although it has been suggested that the beneficial effects of phenobarbital or DDT are caused by their induction of glucuronyl transferase in the liver (Remmer, 1965), conclusive evidence still remains to be demonstrated (Kreek and Sleisenger, 1968; Whelton et al., 1968; Marver and Schmid, 1968).

With regard to public health implications, it appears that people with a partial glucuronyl transferase deficiency may not accumulate significant quantities of PCB-like substances. These substances seem capable of inducing microsomal enzymes which lead to their (PCBs) metabolism and excretion. This constitutes a built-in safety feature which should assist these individuals from accumulating PCB-like chemicals in their bodies. In contrast, individuals with the Crigler-Najjar syndrome seem to have an absolute deficiency of glucuronyl transferase and thus do not have such a built-in compensatory safety system. However, it is not known how effective such a system is in those with the partial enzyme deficiency. It should also be noted that substances which require the activity of glucuronyl transferase for metabolism and excretion and which do not affect microsomal enzyme induction would be expected to accumulate in the body of individuals who have any type of deficiency of glucuronyl transferase (Calabrese, 1977).

IMMUNE RESPONSE AND CARCINOGENICITY

An immunological function of the thymus has long been suspected, since it is packed with lymphocytes, but largely discounted because years of study revealed that it did not form antibodies, trap antigens, or contain more than a rare plasma cell. However, the thymus became recognized as an immunologic agent when mice that had been thymectomized at birth were found to develop a coherent variety of defects (e.g., reduced number of blood lymphocytes, depleted T-cell areas in lymph nodes and spleen, and reduced ability to reject allografts) (Eisen, 1974).

In numerous species, the thymus is fully developed at birth. It atrophies gradually with time. Removal of the thymus from an adult has formerly been considered to produce no obvious effects, because peripheral tissues are already fully populated with many diversified T-cells and T-cell precursors in adaptive transfer (Eisen, 1974).

Directly associated with the size and activity of the thymus is the functionality of cell-mediated immunity (CMI). Numerous investigators believe that the decrease in thymus activity and CMI with aging is causally related to the increased incidence of infections, auto-immune disease, and cancer that often accompany aging (Goldstein et al., 1974). Figure 3 illustrates the maturation pathways of the principal cells (T-cells, B-cells, and macrophages) in the immune response, and Table 3 represents a comparison of functions between T- and B-cells.

Thymosin, a hormone derived from the thymus, has been found to play a crucial role in CMI (White and Goldstein, 1971; Goldstein et al., 1975; Goldstein and White, 1971; Hardy et al., 1971). In fact, injected thymosin has been found to reconstitute cell-mediated immunological responses partially in thymectomized rats. Thymosin treatment for human cancer patients is now being tried on an experimental basis (Goldstein et al., 1975; Marx, 1975). Figure 4 shows the relationship among thymus size, thymosin levels in the blood, and the incidence of cancer and infectious diseases in the human population (Goldstein, 1975).

From an evolutionary perspective, the immune system has been thought to have arisen, as a result of proper mutations and natural selection, to assist the body in resisting infections and in attacking naturally or "artificially" induced tumor growth (Eisen, 1974). Klein (1976) has indicated that immune surveillance against neoplasia is quite efficient for most of the known virus-induced tumors. In fact, the activity of the immune system is especially effective in pre-

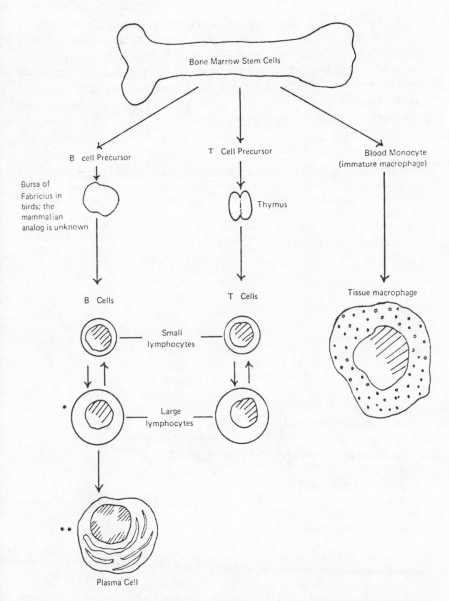

Figure 3. Maturation pathways of the principal cells in the immune response. Antibody molecules are secreted by large lymphocytes (*) and especially by plasma cells (**) in the B-cell lineage. Lymphocytes of B and T lineages are morphologically indistinguishable. *Source.* Eisen (1974).

venting tumor development in animals when a ubiquitous virus is considered (e.g., polyoma in mice, EBV in man) but less effective when the virus is not so widespread (e.g., feline leukemia virus in cats). With regard to immune activity against chemically induced tumors in old animals, the effectiveness is quite weak. Klein (1976) suggests that this may be due to the absence of preselection for appropriate immune responsiveness genes during evolution.

Table 3. Comparison of Functions between Mouse B and T Lymphocytes

Properties	B Cells	T Cells
Differentiation (from uncommitted Ag-insensitive "stem" cells to Ag-sensitive cells) in:	Bursa of Fabricius (in birds) or as yet unknown equivalent in mammals	Thymus
Approximate frequency (%) in:		
Blood	15	85
Lymph (thoracic duct)	10	90
Lymph node	15	85
Spleen	35	65
Bone marrow	Abundant	Abundant
Functions		
Secretion of antibody molecules	Yes (large lymphocytes and plasma cells)	No
Helper function (react with "carrier" moieties of the immunogen)	No	Yes
Effector cell for cell-mediated immunity	No	Yes
Distribution in lymph nodes and spleen:	Clustered in follicles around germinal centers	In inter-follicular areas
Susceptibility to inactivation by:		
X-Irradiation	++++	+
Corticosteroids	++	+
Antilymphocytic serum (ALS)	+	++++

Source. Eisen (1974).

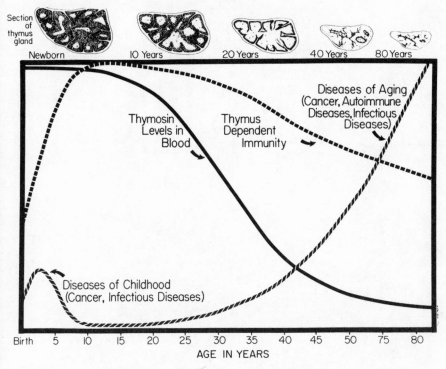

Figure 4. Thymus-dependent senescence of immunity in man. *Source.* Goldstein et al. (1975).

At this point, it is important to mention that several well-known carcinogens such as long-fibered asbestos and arsenic are thought to have "latent" periods of possibly 30 years before influencing cancer growth in the exposed individual (Table 4). Based on the knowledge of thymus gland activity and its various statistical associations with human disease, it seems possible that the latent period of known carcinogens may be a function of the body's CMI. Consequently, it is possible that such substances may not act with a latency period at all. That is, asbestos-induced tumors could be constantly produced, but an active functioning CMI may be suppressing the tumor growth; however, at a time when the CMI sufficiently decreases in efficiency, a cancer may develop to such an extent as to cause death or require surgery. Studies by Kagan et al. (1976) indicated that workers exhibiting either pleural thickening or parenchymal asbestosis along with a definite history of occupational exposure to asbestos exhibited significantly

Table 4. Classification of Occupational Carcinogens

Agents	Affected Organs	Incubation Period (Years)	Risk Ratio	Occupation
Organic agents				
Aromatic hydrocarbons				
Coal soot, coal tar, other products of coal combustion	Lung, larynx, skin, scrotum, urinary bladder	9–23	2–6	Gashouse workers, stokers, and producers; asphalt, coal tar, and pitch workers; coke-oven workers; miners; still cleaners; chimney sweeps
Petroleum, petroleum coke, wax, creosote, anthracene, paraffin, shale, mineral oils	Nasal cavity, larynx, lung, skin, scrotum	12–30	2–4	Contact with lubricating, cooling, paraffin or wax fuel oils, or coke; rubber fillers; retortmen; textile weavers; diesel jet testers
Benzene	Bone marrow (leukemia)	6–14	2–3	Explosives, benzene, or rubber cement workers; distillers; dye users; painters; shoemakers
Auramine, benzidine, α-naphthylamine, β-naphthylamine, magenta, 4-aminodiphenyl, 4-nitrodiphenyl	Urinary bladder	13–30	2–90	Dyestuffs manufacturers and users; rubber workers (pressmen, filtermen, laborers); textile dyers; paint manufacturers

Agent	Site			Workers
Alkylating agents				
Mustard gas	Larynx, lung, trachea, bronchi	10–25	2–36	Mustard gas workers
Others				
Isopropyl oil	Nasal cavity	10+	21	Producers
Vinyl chloride	Liver (angiosarcoma), brain	20–30	200 (liver) 4 (brain)	Plastic workers
Bis(chloromethyl) ether, chloromethyl methyl ether	Lung (oat cell carcinoma)	5+	7–45	Chemical workers
Inorganic agents				
Metals				
Arsenic	Skin, lung, liver	10+	3–8	Miners; smelters; insecticide makers and sprayers; tanners; chemical workers; oil refiners; vintners
Chromium	Nasal cavity and sinuses	15–25	3–40	Producers, processors, and users; acetylene and aniline workers; bleachers; glass, pottery, and linoleum workers; battery makers

Table 4 (Continued)

Agents	Affected Organs	Incubation Period (Years)	Risk Ratio	Occupation
Inorganic agents—Metals				
Iron oxide	Lung, larynx	—	2–5	Iron ore (hematite) miners; metal grinders and polishers; silver finishers; iron foundry workers
Nickel	Nasal sinuses, lung	3–30	5–10 (lung) 100+ (nasal sinuses)	Nickel smelters, mixers, and roasters; electrolysis workers
Fiber				
Asbestos	Lung, pleural and peritoneal mesothelioma	4–50	1.5–12	Miners; millers; textile, insulation, and shipyard workers
Dusts				
Wood	Nasal cavity and sinuses	30–40	—	Woodworkers
Leather	Nasal cavity and, sinuses, urinary bladder	40–50	50 (nasal sinuses) 2.5 (bladder)	Leather and shoe workers

Physical agents

Nonionizing radiation

Ultraviolet rays	Skin	Varies with skin pigment and texture	—	Farmers; sailors

Ionizing radiation

X-Rays	Skin, bone marrow (leukemia)	10–25	3–9	Radiologists; medical personnel
Uranium, radon, radium, mesothorium	Skin, lung, bone, bone marrow (leukemia)	10–15	3–10	Radiologists; miners; radium dial painters; radium chemists

Other

Hypoxia	Bone	—	—	Caisson workers

Source. Cole and Goldman (1975).

23

reduced levels of IgA, IgG, IgM, and IgE as well as the T-lymphocytes. Also, a high occurrence of cold-reactive serum lymphocytotoxins was noted in these asbestos-exposed workers as compared with the control group. Such evidence indicates an association between asbestos exposure and the immune surveillance system and may indicate a possible role of asbestos in the induction of altered immune function.

The administration of immunosuppressive drugs with organ transplantation has often led to the appearance of cancer. This implies that the normal immunologic surveillance and control of "hosts" of incipient tumors have been interrupted (Fahey, 1971). Thus, chemicals which depress the immune system should be considered as possible cofactors in carcinogenesis. It is important to note that various carcinogens, including some of the polycyclic aromatic hydrocarbons, are known to depress the activity of the immune system (Szakal and Hanna, 1972; Stjernswärd, 1966, 1969).

PREGNANCY: A CONDITION INFLUENCING
THE TOXICITY OF POLLUTANTS

Pregnancy marks a time of numerous physiological changes in women. For example, the concentration of red blood cells in the blood falls, since the increase of plasma volume is greater than the increase of red blood cell volume. Also, blood protein (a, β, λ globulins) concentration is significantly affected. Levels of a_1-antitrypsin and a_2-macroglobulin are raised during pregnancy. Nutritional requirements are markedly affected, especially for iron and calcium (Hunt, 1975). Of a group of "normal" pregnant women who did not receive supplemental iron, 85 percent had no iron stores (Monsen et al., 1967). Other researchers have reported anemia in 15 to 58 percent of pregnant women (de Leeuw et al., 1966).

As a result of the numerous physiological modifications which the pregnant woman must experience, there exists the possibility that she will respond differently to environmental stressors as compared with the nonpregnant woman. Since deficiencies in iron are known to predispose individuals to the toxic effects of Cd, Mn, and Pb, pregnant women would be at considerably high risk (see pp. 106, 109, and 112). An investigation concerning the effects of occupational exposure of pregnant women to Cd (Tsuetkova, 1970) revealed that the children of women exposed during pregnancy weighed significantly less than the controls.

The National Institute of Occupational Safety and Health (NIOSH) Criteria Document concerning beryllium (Br) states that the delayed onset of pneumonitis is frequently precipitated by some acute stress, including surgery, viral respiratory infections, and pregnancy (Hunt, 1975). Other authors have noted in their study that 40 percent of the women with chronic diseases who became pregnant following Br exposure developed pneumonic symptoms along with the pregnancy (Hardy and Stoeckle, 1959). It has been further reported that of the 95 dead females in the Beryllium Registry, 63 of them were noted as having pregnancy as a probable precipitating factor (Hall et al., 1959).

Endogenous CO production is known to fluctuate considerably during pregnancy (Hunt, 1975). For example, Linderholm and Lundstrom (1969) have noted an increase of 50 percent. This may result in part from the increase in red blood cell mass during pregnancy and the CO production by the fetus. According to Curtis et al. (1955), over 20 cases of CO poisoning during pregnancy have been reported. In the 10 cases where the mother survived, 8 of the neonates developed neurological sequelae, and 3 of them died later, with evidence of brain damage at autopsy.

Organochlorine compounds may adversely affect female reproductive physiology. For example, DDT exposure to female workers affected ovarian-menstrual function by being frequently associated with the onset of menorrhagia and dysmenorrhea. Disruption in childbearing such as spontaneous miscarriages, pregnancy toxicosis, and premature bursting of the amniotic sac also occurred significantly more often in exposed workers as compared with individuals in the control group (Veis, 1970).

Hanhijärvi (1974) investigated the effect of pregnancy on free ionized plasma fluoride concentrations in women from an artificially fluoridated drinking water community (1 ppm). The results indicated that the plasma-ionized fluoride concentration decreases steadily and significantly during pregnancy, reaching the lowest levels in the later stages. The author concluded that the low levels of plasma-ionized fluoride result from the ability of the fetus to incorporate increasing amounts of fluoride in the mineralizing tissues. The health significance of this biochemical change remains to be investigated. Table 5 summarizes the pollutants which cause enhanced risk to women as a result of the physiological modifications of pregnancy.

Table 5.　Pollutant Toxicity Effected by Physiological Modifications of Pregnancy

Pollutants Causing Increased Risk	Physiological Modifications
Cadmium, lead, manganese	Increased dietary requirements of calcium and iron
Carbon monoxide (CO)	Excessive endogenous production of CO
Organophosphate insecticides	Hormonal alterations
Beryllium	Respiratory disease susceptibility

SUSCEPTIBILITY TO ENVIRONMENTAL STRESSES AS AFFECTED BY CIRCADIAN RHYTHMS

Numerous body functions undergo variations recurring at about 24-hr intervals in the presence or absence of known environmental changes with similar periods. This applies to continuous but rhythmic phenomena, with a peak and trough repeating itself every 24 hr, as well as to discrete events, occurring about once each day. The time intervals separating these consecutive periodic events are similar but often not identical. Such periods are called circadian, meaning "about a day" (Luce, 1970). Other well-known periodicities are specific for a given organ system, organ, or tissue. Circadian rhythms are more generally present in metabolizing structures at various levels of physiological organization (Halberg, 1960).

The brain exhibits fast rhythms, with several cycles each second, the heart and respiratory system show many cycles per minute, and the human ovary demonstrates approximately one cycle a month; yet, in all these organs, some functions indicate circadian rhythm. Fast or slow, organ-specific rhythms have to be coordinated with metabolism, and rhythmic events have to be integrated with each other and the environment. The body achieves such an integration by developing "sequential order in time." The circadian organization of interacting body functions is instrumental in assisting the body in its temporal adaptation to our earthly environment (Halberg, 1960).

Circadian rhythms have been reported in cell growth, mitosis, hormonal levels, body temperature, central nervous system activity, and so on. Susceptibility to various agents has also been shown to have a circadian rhythm (Luce, 1970; Halberg, 1960) (Figures 5 and 6).

Circadian System of the Mouse

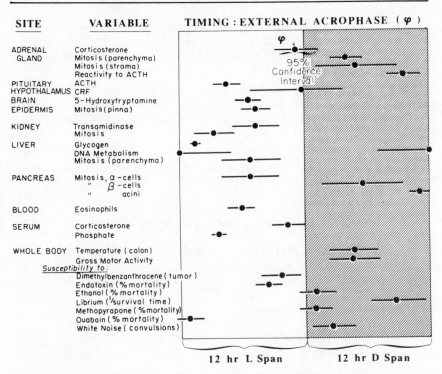

Figure 5. Circadian system of the mouse. *Source.* Luce (1970).

C mice exhibited a susceptibility rhythm to *E. coli* endotoxin. Susceptibility varied predictably and significantly along a 24-hr time period. A concentration of endotoxin, which would be nonlethal to most animals when given in the middle of daily dark periods, is highly lethal when given 8 to 12 hr earlier or later (Halberg et al., 1960). Similar experiments with C mice also established a 24-hr rhythm in susceptibility to toxic doses of ethanol (Hans and Halberg, 1959; Hans et al., 1959). Using I mice, Halberg et al. (1955) reported 24-hr periodicities with respect to susceptibility to audiogenic convulsion and ability to recover from convulsions. In one group of young mice, the convulsive risk was 11 percent by day and 63 percent by night and in another, 0 percent by day and 85 percent by night. On the human level, epileptologists have recognized the unequal "within day" (i.e., circadian) distribution of seizures in some patients and have investigated convulsive periodicity as a possible clue to understanding seizure initiation (Halberg et al., 1958).

Carcinogen- (benzpyrene) induced malignancy in mice is also a function of circadian periodicity. It is apparently contingent on the phase relation among rhythms at the time of carcinogen administration (Mottram, 1945).

It has been suggested that the periods of increased susceptibility have their spatial counterpart in certain vulnerable spots of the body. However, circadian peaks in susceptibility to various agents theoretically do not have to be identical (Halberg, 1960).

It becomes apparent that all people have periods throughout each 24-hr period when they, as individuals, are more susceptible to various

Human Circadian System
Birth, Death, Morbidity, Susceptibility And Reactivity

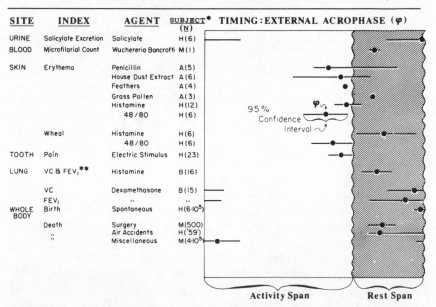

This summary of human susceptibility rhythms was prepared by the Chronobiology Laboratory at the University of Minnesota.
*H= Healthy; M= Morbid; A= Allergic; B= Asthmatic, bronchitic, or emphysematous. ** VC= Vital Capacity; FEV= 1 -second forced expiratory volume. Acrophase (the peak represented by the dot), indicates the circadian phase of maximum response.

Figure 6. Human circadian system: birth, death, morbidity, susceptibility, and reactivity. *Source.* Luce (1970).

environmental stressors. If data from which environmental standards are derived do not consider these "weak links," the adequacy of the standard to protect human health is highly questionable.

PHASE SHIFTS AND SUSCEPTIBILITY
TO ENVIRONMENTAL STRESSES

Several investigators have indicated that social routine constitutes the dominant rhythmic variation in the environment. This rhythm determines the timing of circadian body rhythms when the individual has been in the time zone long enough to be adapted to changes accompanying local clock time. Social activities, in our evolutionary-biological development, are, however, based primarily on astronomically fixed cycles, and the end result is the establishment of a constant time relation between the variation of the dominant external and the biologic internal rhythms (Luce, 1973; Conray and Mills, 1970; Felton and Patterson, 1971). An efficient coordination of internal metabolic adaptations to the external environment is of vital survival value.

Disruptions of this time coordination relationship are known to cause restlessness, nervousness, headache, gastrointestinal irregularities, anorexia, fatigue, slower reaction time, and error proneness (Mott et al., 1965; Strughold, 1952; Kleitman and Jackson, 1950; Mills, 1967). Some authors have suggested that these conditions could lower resistance to disease. Disruptions of this type may take place when workers are scheduled for shifts and assignments which rotate around the clock (Felton and Patterson, 1971).

When environmental conditions are modified by the shifting of work hours, the pattern of the social routine is likewise changed. When individuals change their regular 8-hr shifts and go abruptly to new work hours, it takes time for their body rhythms to make the adjustment, and they may not adjust at all (Felton and Patterson, 1971). Several reports have indicated that individuals vary in their ability to adjust to a different sleep-wake cycle—2 to 3 days to 2 weeks (Felton and Patterson, 1971; Mott et al., 1965; Mills, 1967). Teleky (1943) also found that night workers in certain industries never successfully adjusted their temperature cycles.

Although research has not yet established definitive relationships linking phase shifting and disruption in the circadian rhythms to increased susceptibility to environmental stressors, this definitely appears to be an area of potential concern. Any time an efficient system is disrupted, the reliability of many important functions becomes suspect. Since work shifts are used by occupations including industry, medicine, police, and firemen, millions of people are being exposed to such physiological disruptions. Therefore, at least theoretically, these millions of people may possibly represent a major high-risk group to various environmental stressors.

GASTROINTESTINAL ABSORPTION OF POLLUTANTS AS AFFECTED BY AGE

One of the main avenues of exposure to various pollutants such as the heavy metals is via the gastrointestinal tract. Animal studies by Forbes and Reina (1972) and Taylor et al. (1962) and human studies by Schulz and Smith (1958) have indicated that absorption rates are significantly higher in the very young as compared with any other age group (Figure 7). Forbes and Reina (1972) have shown that age-absorption curves for strontium (Sr) and Pb are similar to that for Fe in the rat.

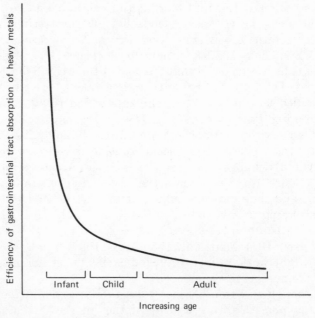

Figure 7.   The relationship of age to the efficiency of gastrointestinal tract absorption of heavy metals.

The age effect on gastrointestinal absorption is shared by a number of multivalent cations which are readily absorbed by the infant rat but rejected in large part by the adult animal. Forbes and Reina (1972) stated that a mechanism is apparently acquired during growth which aids the adult in avoiding undue accumulation of certain elements normally present in the diet; the infant, however, appears to absorb these identical elements readily.

FLUORIDE RETENTION AS A FUNCTION OF AGE

The renal fluoride clearance and daily excretion of the fluoride ion increase slightly up to the age of 50 in individuals from both fluoridated and nonfluoridated areas, after which the renal fluoride clearance decreases (Hanhijärvi, 1974). One proposed explanation for the increase of fluoride clearance with age could be the decreasing accumulation of fluoride in the bones. Up to the age of 50 the bones become progressively saturated with fluoride; this would increase the renal fluoride clearance, since less fluoride is then retained in the bones. A possible reason why renal fluoride clearance decreases after the age of 50 is that renal function weakens with old age (Hanhijärvi, 1974). Such information indicates that people over 50 years old are clearly predisposed to retaining higher levels of fluoride than other age groups and thus may represent a group at increased risk with respect to the potential toxic effects caused by excessive fluoride retention. Evidence of any specific adverse health response to this additional fluoride exposure remains to be investigated.

AGE, RADIATION SENSITIVITY, AND CANCER

The toxic effects of ionizing radiation on human beings are well known. Initial human epidemiologic studies indicated that the occurrence of leukemia is greater among individuals exposed to relatively large doses of ionizing radiation as compared with the general public (United Nations Scientific Committee on the Effects of Atomic Radiation, 1964). The exposed individuals considered in the United Nations report included Japanese survivors of the atomic bombing at Hiroshima and Nagasaki (single doses of 100 to 900 rads), patients irradiated for ankylosing spondylitis (100 to 300 rads for 1 month), radiologists exposed to ionizing radiation (considerably greater than 100 rads), and children irradiated *in utero* in the course of pelvic X-ray examinations (about 3 rads).

Later studies of the Japanese survivors demonstrated that excess numbers of thyroid gland cancer, breast cancer, and lung cancer also occurred, in addition to the leukemias previously recognized (Tamplin and Gofman, 1970). From the research of Court-Brown and Doll (1965) on 14,000 people receiving therapeutic radiation for the arthritislike disorder, rheumatoid spondylitis, it was also determined that numerous types of cancer in organs (lungs, stomach, lymphatic and blood-forming organs, pancreas, pharynx, bone, colon, and a variety of additional cancers of miscellaneous origin) heavily exposed to irradiation occurred in addition to the expected increase in leukemias.

Additional analysis of the available information concerning radiation-induced human cancer indicated that, although 50 rads were required to double the spontaneous cancer incidence in adults, only approximately 5 to 10 rads were required in children, thus indicating a marked increase in the sensitivity of children to the toxic effects of radiation (Tamplin and Gofman, 1970). Stewart et al. (1958), Stewart and Kneale (1968, 1970), MacMahon (1962), and MacMahon and Hutchison (1964) have further indicated that *in utero* infants are even at higher risk. They showed that all types of childhood cancer and leukemia are doubled by extremely small dosages of radiation. Figure 8 indicates how the toxic effects of radiation are clearly effected by age (Tamplin and Gofman, 1970).

Stewart and Kneale (1970) specifically indicated that $1\frac{1}{2}$ rads doubled the frequency of leukemia in children if X-rays were taken in the latter half of pregnancy. However, if X-rays were taken during the first trimester of pregnancy only $\frac{1}{3}$ rad was needed to double the incidence of cancer in the first 10 years of life.

Graham (1972) has suggested that there may be even more highly susceptible subgroups among children. The subgroups seem to have a predisposition toward developing leukemia after X-ray exposures that have no obvious effect on others. Diamond et al. (1973) have also suggested that X-ray sensitivity is a function of racial ancestry. For example, preliminary epidemiological studies imply that white children may be 40 to 60 percent more susceptible than black children to the harmful effects of prenatal radiation exposure. Thus, of the prenatally exposed white children, mortality during the first 10 years of life was almost twice that of the white controls (nonexposed) for all causes of death except cancer (not including leukemia), congenital malformations among males, and nervous system diseases of females. With

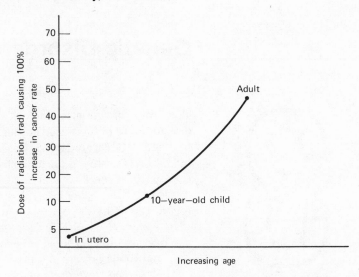

Figure 8.   The effects of age and radiation dose on the spontaneous rate of cancer incidence.
Source.   Tamplin and Gofman, 1970.

regard to leukemia, the incidence of death was approximately three times higher among the exposed than the controls. However, in blacks the mortality of the exposed children was not significantly different from the control group.

In the first community epidemiological studies of humans exposed to high levels of radium-226 (4 pCi/1) in drinking water as compared with controls, all age groups in the treatment group experienced greater numbers of bone cancers than their equivalent control groups. However, the increases in bone cancer became even more pronounced in the under 30-year-old group. Such information confirms the previously discussed research which indicated that radiation sensitivity is markedly affected by age (Peterson et al., 1966).

Genetic Disorders

The role of genetic factors in the development of increased susceptibility to the toxic or carcinogenic effects of pollutants has assumed an important posture with respect to standard setting (Stokinger and Scheel, 1973). In fact, a recent report by the National Academy of Sciences (NAS) (1975) devoted a special section on the role of inborn (genetic) metabolic errors as predisposing factors in the development of toxicity from occupational and environmental pollutants. In their section on genetic disorders, the NAS report (1975) listed 92 human genetic disorders which were adapted from McKusick (1970). Table 6 lists these 92 genetic disorders and indicates which of the 92 are thought to predispose the affected individuals to the toxic effects of pollutants.* This chapter examines the most significant genetic disorders associated with the exacerbation of pollutant toxicity and in so doing attempts to present the actual or theoretical basis for such an association.

---

* By 1972, 150 genetic diseases for humans were identified in the third edition of *The Metabolic Basis of Inherited Disease* (Stanbury et al., 1972).

RED BLOOD CELL DISORDERS

Hereditary Blood Disorders and Hemolytic Susceptibility

There is a broad group of genetic diseases that result in either producing or predisposing afflicted individuals to the development of hemolytic anemias. For example, these diseases include conditions in which (1) abnormal hemoglobin occurs (sickle cell), (2) inability to manufacture one or another of the peptide globin chains of hemoglobin occurs (thalassemia), and (3) particular enzyme deficiencies of the Embden-Meyerhoff or hexose monophosphate shunt metabolic pathways—glucose-6-phosphate dehydrogenase (G-6-PD)—as well as other cellular enzyme deficiencies (e.g., catalase) may occur.

Table 6. Disorders in Which a Deficient Activity of a Specific Enzyme Has Been Demonstrated in Man

Condition	Enzyme with Deficient Activity
Acatalasia[a]	Catalase
Acid phosphatase deficiency	Lysosomal acid phosphatase
Adrenal hyperplasia I	21-Hydroxylase[b]
Adrenal hyperplasia II	11-Beta-hydroxylase[b]
Adrenal hyperplasia III	3-Beta-hydroxysteroid dehydrogenase[b]
Adrenal hyperplasia V	17-Hydroxylase[b]
Albinism[a]	Tyrosinase
Aldosterone synthesis, defect in	18-Hydroxylase[b]
Alkaptonuria	Homogentisic acid oxidase
Angiokeratoma, diffuse (Fabry)	Ceramidetrihexosidase
Apnea, drug-induced[a]	Pseudocholinesterase
Argininemia	Arginase
Argininosuccinic aciduria	Argininosuccinase
Aspartylglycosaminuria	Specific hydrolase (AADC-ase)
Carnosinemia	Carnosinase
Cholesterol ester deficiency (Norum's disease)	Lecithin cholesterol acetyltransferase (LCAT)
Citrullinemia	Arginosuccinic acid synthetase
Crigler-Najjar syndrome[a]	Glucuronyl transferase
Cystathioninuria	Cystathionase
Formininotransferase deficiency	Formininotransferase
Fructose intolerance	Fructose-1-phosphate aldolase

Table 6 (Continued)

Condition	Enzyme with Deficient Activity
Fructosuria	Hepatic fructokinase [b]
Fucosidosis	Fucosidase
Galactokinase deficiency	Galactokinase
Galactosemia	Galactose-1-phosphate uridyl transferase
Gangliosidosis, generalized	β-Galactosidase
Gaucher's disease	Glucocerebrosidase
G-6-PD deficiency (favism, primaquine sensitivity, nonspherocytic hemolytic-anemia) [a]	Glucose-6-phosphate dehydrogenase
Glycogen storage disease I	Glucose-6-phosphatase
Glycogen storage disease II	Alpha-1-4-glucosidase
Glycogen storage disease III	Amylo-1-6-glucosidase
Glycogen storage disease IV	Amylo (1-4 to 1-6) transglucosidase
Glycogen storage disease V	Muscle phosphorylase
Glycogen storage disease VI	Liver phosphorylase [b]
Glycogen storage disease VII	Muscle phosphofructokinase
Glycogen storage disease VIII	Liver phosphorylase kinase
Gout, primary (one form)	Hypoxanthine guanine phosphoribosyl transferase
Hemolytic anemia	Diphosphoglycerate mutase
Hemolytic anemia [a]	Glutathione peroxidase
Hemolytic anemia [a]	Glutathione reductase
Hemolytic anemia [a]	Hexokinase
Hemolytic anemia	Hexosephosphate isomerase
Hemolytic anemia	Phosphoglycerate kinase
Hemolytic anemia	Pyruvate kinase
Hemolytic anemia	Triosephosphate isomerase
Histidinemia	Histidase
Homocystinuria	Cystathione synthetase
Hydroxyprolinemia	Hydroxyproline oxidase
Hyperammonemia I	Ornithine transcarbamylase
Hyperammonemia II	Carbamyl phosphate synthetase
Hyperglycinemia, ketotic form	Propionate carboxylase [b]
Hyperlysinemia	Lysine-ketoglutarate reductase
Hyperoxaluria I glycolic aciduria	2-Oxo-glutarate-glyoxylate carboligase
II Glyceric aciduria	D-Glyceric dehydrogenase
Hyperprolinemia I	Proline oxidase deficiency
Hyperprolinemia II	δ-1-Pyrroline-5-carboxylate dehydrogenase
Hypophosphatasia	Alkaline phosphatase

Table 6 (Continued)

Condition	Enzyme with Deficient Activity
Isovalericacidemia	Isovaleric acid CoA dehydrogenase
Lactase deficiency, adult, intestinal	Lactase
Lactose intolerance of infancy	Lactase
Leigh's necrotizing encephalo-myelopathy	Pyruvate carboxylase
Lesch-Nyhan syndrome	Hypoxanthine-guanine phosphoribosyl-transferase
Lipase deficiency, congenital	Lipase (pancreatic)
Lysine intolerance	L-Lysine: NAD-oxido-reductase
Mannosidosis	a-Mannosidase
Maple sugar urine disease	Keto acid decarboxylase
Metachromatic leukodystrophy	Arylsulfatase A (sulfatide sulfatase)
Methemoglobinemia[a]	NADH-methemoglobin reductase
Methylmalonicaciduria	Methylmalonyl-CoA carboxymutase
Myeloperoxidase deficiency with disseminated candidiasis	Myeloperoxidase
Niemann-Pick disease	Sphingomyelinase
Oroticaciduria	Orotidylic pyrophosphorylase orotidylic decarboxylase
Phenylketonuria[a]	Phenylalanine hydroxylase
Porphyria, congenital erythro-poietic[a]	Uroporphyrinogen III cosynthetase
Pulmonary emphysema[a]	a_1-Antitrypsin
Pyridoxine-dependent	Glutamic acid decarboxylase[b]
Pyridoxine-responsive anemia	δ-Aminolevulinic acid synthetase[b]
Refsum's disease	Phytanic acid a-oxidase
Sarcosinemia	Sarcosine dehydrogenase[b]
Sucrose intolerance	Sucrose, isomaltase
Sulfite oxidase deficiency[a]	Sulfite oxidase
Tay-Sachs disease	Hexosaminidase A
Testicular feminization	Δ^4-5a-Reductase[b]
Thyroid hormonogenesis, defect in	Iodothyrosine dehalogenase (deiodinase)
Trypsinogen deficiency disease	Trypsinogen
Tyrosinemia I[a]	Para-hydroxyphenylpyruvate oxidase
Tyrosinemia II[a]	Tyrosine transaminase
Valinemia	Valine transaminase
Vitamin D–resistant rickets	Cholecalciferase[b]
Wolman's disease	Acid lipase
Xanthinuria	Xanthine oxidase

Table 6 (Continued)

Condition	Enzyme with Deficient Activity
Xanthurenic aciduria	Kynureninase
Xeroderma pigmentosa[a]	Ultraviolet specific endonuclease[b]

[a] Associated with the exacerbation of toxic effects of at least one pollutant (see text).

[b] In some conditions, as well as some that are not listed, deficiency of a particular enzyme is suspected but has not been proved by direct study of enzyme activity.

Source. Adapted from McKusick (1970) and NAS (1975). Reprinted with permission of the National Academy of Sciences. National Research Council-National Academy of Sciences-National Academy of Engineering. 1975. *Principles for Evaluating Chemicals in the Environment.* National Academy of Sciences, Washington, D.C., pp. 338–340.

The state of knowledge concerning these diseases is quite variable, with considerable information known about sickle-cell disease, catalase and G-6-PD deficiencies, and thalassemia, but little is known about most of the enzyme deficiencies of the glycolytic pathway. Special emphasis is given here to sickle-cell disease, catalase and G-6-PD deficiencies, and thalassemia, since these are the most frequently occurring hereditary blood diseases and, as previously stated, about which most is known. However, a brief consideration of the less frequently occurring and less understood enzyme deficiency conditions associated with hemolytic anemias is provided. At this time, it must be emphasized that, although it is suggested that individuals with such genetic deficiencies may be at increased risk to the action of some hemolytic agents, research in this area is generally lacking. However, because of their theoretical potential for causing hemolysis under particular "oxidant" stress conditions and for the sake of an overall perspective in this area, the following summary of glycolytic enzyme deficiencies is provided in Table 7.

In order to facilitate an understanding of subsequent sections concerning several of the hereditary blood disorders, a brief review of how the normal red blood cell as well as the hemoglobin molecule functions follows.

Normal Red Blood Cell Functioning. The developing red blood cell has the complete metabolic potential for replication, differentiation, and maintenance. However, on emergence into the reticulocyte stage,

Table 7. Enzyme Deficiencies Associated with Hemolysis

Enzyme Deficiency	Frequency	Genetics
Embden-Meyerhoff pathway:		
Hexokinase (HK)	Rare	Most likely autosomal recessive
Glucose phosphate isomerase (GPI)	Rare	Autosomal recessive
Phosphofructokinase	Rare	
Triosephosphate isomerase (TPI)	Rare	Autosomal recessive
2,3-Diphosphoglyceromutase (2,3-DPGM)	Rare	Thought to be autosomal recessive
Phosphoglycerate kinase (PGK)	Rare	x-Chromosome linkage
Pyruvate kinase (PK)	Most frequent next to G-6-PD	Autosomal recessive
Hexose monophosphate (HMP) shunt pathway:		
Glucose-6-phosphate dehydrogenase (G-6-PD)	Common	x-Chromosome
6-Phosphogluconate dehydrogenase (6-PGD)	Rare	Autosomal recessive
Nonglycolytic pathway:		
Glutathione reductase[a] (GSSG-R)	Uncertain	Uncertain
Glutathione peroxidase (GSH-Px)	Rare	Uncertain
Glutathione synthetase[b] reaction II	Rare	Thought to be autosomal recessive
ATPase	Unknown	Unknown

[a] Levels of GSSG-R are known to be influenced by riboflavin and flavin-adenine dinucleotide (FAD) in the diet (Beutler, 1969).

[b] Individuals with this deficiency are known to be sensitive to "oxidant-type" drugs such as primaquine (Carson et al., 1961).

Source. Valentine and Tanaka (1972). From *The Metabolic Basis of Inherited Disease,* Stanbury, J. B., Wyngaarden, J. B., and Fredrickson, D. S. (Eds.). Copyright 1972, McGraw-Hill Book Company. Used with permission of McGraw-Hill Book Company.

the red blood cell becomes nonnucleated and has only limited capacity for protein and lipid synthesis and the incorporation of iron into hemoglobin. The next stage, or the mature red blood cell stage, is also nonnucleated and has no DNA, RNA, mitochondria, or other intracellular organelles and only relatively nonfunctional remnants of Krebs cycle metabolism. The mature red blood cell can manufacture no additional protein and thus no new enzymes. The catalytic proteins each have their own biologic half-life and decay at different rates as the cell ages. However, the mature red blood cell synthesizes certain compounds such as glutathione (GSH), nicotinamide dinucleotide (NAD), flavinadenine dinucleotide (FAD), and adenosine triphosphate (ATP) (Valentine and Tanaka, 1972).

The main function of the immature red blood cell is to produce hemoglobin. The immature red cell becomes "mature" just shortly after entering the circulatory system. The mature red cell functions to transport oxygen and carbon dioxide. To carry out these activities, it is necessary to have functional hemoglobin molecules and structural integrity of the cell. A functional hemoglobin molecule requires an intracellular reducing mechanism to ensure the constant reduction of methemoglobin (MetHb) to hemoglobin. The mechanism involves the regeneration of NADH and NADPH. To maintain the structural integrity of the cell, it is necessary to keep the proper electrolyte gradient across the cell membrane, and this depends on the availability of energy or ATP (Valentine and Tanaka, 1972; Friedman, 1971).

The volume and thickness of the red blood cell rely on the control of ion transport across the cell membrane. To maintain the proper volume and osmotic balance in light of the osmotic pressure by nondiffusable cellular constituents, including hemoglobin and some glycolytic intermediates, the red cell must carry on active transport of cations. For example, sodium is transported into the cell "passively" along an electrochemical gradient and is subsequently "pushed" out of the cell via the expenditure of cellular energy. If glycolysis is prevented and if high-energy phosphate molecules (ATP) are not produced, or if the rate of passive entry exceeds the capacity of active transport to balance the flow, the vital regulation of cations and thus membrane stability are disrupted (Jandl and Cooper, 1972).

The red cell needs a source of energy production for synthesis activity, maintenance of proper ion concentration gradients, and the reduction of MetHb, which is constantly formed. Because the mature red blood cell has a reduced capacity to use oxygen and is unable to use pyruvate, glucose is primarily metabolized via the anaerobic (Embden-Meyerhoff) pathway, and lactic acid accumulates. Besides the Embden-Meyerhoff pathway, the mature red cell also has an alternative, oxidative pathway, the hexose monophosphate shunt (HMP) or pentose phosphate pathway. The HMP involves glucose-6-phosphate being converted in most cases to fructose-6-phosphate. During the intermediate steps, several important events occur during which NADP is involved as a hydrogen donor to function in the reduction of oxidized glutathione (GSSG). This particular scheme is necessary for the maintenance of cell membrane integrity. When the HMP is stimulated, for every three glucose molecules which travel the HMP route, three molecules of fructose-6-P are formed. Generally, the HMP is not very active and usually oxidizes less than 10 percent of the glucose. The relative activity of the shunt is governed by the availability of NADP, and this is likewise dependent on the oxidation-reduction state of glutathione (Valentine and Tanaka, 1972).

Red blood cell survival depends on preserving a smooth biconcave shape, maintaining free suspendibility and the proper membrane surface area to permit the cell to undergo the extremes of cell deformation which constant circulation may cause. Slight changes in the cell's surface area–volume ratio or in the stickiness of its surface may add to the trapping and ultimate death and destruction of the cell (Jandl and Cooper, 1972).

Background Information Concerning Hemoglobin. There are two types of hemoglobin: fetal and adult. These are designated by the symbols Hb F and Hb A, respectively. At birth 60 to 90 percent of the hemoglobin is Hb F. Usually by 4 months of age, only trace amounts remain except in Negro children, where it takes somewhat longer to disappear. Differences between Hb F and Hb A occur with regard to solubility, ultraviolet spectrum, alkali denaturation, oxygen affinity, and antigenic activity. Hemoglobin F is genetically transmitted independently of adult hemoglobin variants (Lehmann and Huntsman, 1972).

Adult hemoglobin consists of at least two components: Hb A, the major component, and Hb A_2, which makes up about 2 percent of the total.

In human hemoglobin, each molecule contains four polypeptide chains of which two, called a, are common to Hb A, F, and A_2. The second pair consists of β, γ, or δ chains and differs in each of these three hemoglobins:

$$\text{Hb A} = a_2\beta_2$$
$$\text{Hb F} = a_2\gamma_2$$
$$\text{Hb A}_2 = a_2\delta_2$$

The a chain has 141 amino acid residues, and the other chains consist of 146 each. The amino acid sequences have been determined for each chain.

More than 100 variant hemoglobins have been identified. Specific designations have been given to many of these variant hemoglobin molecules, for example, Hb S, Hb C, Hb D, Hb E, and Hb M. With regard to Hb S, it differs from Hb A in the amino acid sequence of the poly-peptide chains which form the globin molecule. Such structural modi-fications lead to the development of the hemoglobin molecule respons-ible for sickle-cell trait (Lehmann and Huntsman, 1972).

Sickle-Cell Trait and Sickle-Cell Anemia

Sickle-cell trait is due to the presence of an abnormal hemoglobin, Hb S, in the red blood cells of certain individuals. It is the Hb S which causes the development of the sicklelike shape of the red blood cell. Hemo-globin S is less soluble than the normal hemoglobin molecule when deoxygenated (Lehmann and Huntsman, 1972). For example, research by Perutz and Mitchison (1950) has indicated that oxygenated Hb A and Hb S have similar solubility. However, when they are deoxygenated, the solubility of Hb A falls about 50 percent as compared with 5000 percent for Hb S. This creates the formation of a gel structure in the red blood cells, with a resulting distortion of the rounded cell shape to the form of a sickle. This modification of cell shape increases the blood viscosity (Greenberg et al., 1957). The increase in blood viscosity slows down the blood flow, which further decreases oxygenation and thus leads more rapidly to sickle-cell formation. Furthermore, if sickle cells are held in a deoxygenated state for more than 15 sec, sickling is initiated, and this, of course, may be enhanced by increased blood viscosity (Alli-son, 1956). Sickle-cell anemia results when an individual is homozygous with respect to the abnormal gene. If an individual is heterozyous for the Hb S gene, he has the sickle-cell trait.

The homozygous individual with sickle-cell anemia has the Hb S molecule in most of the red blood cells; those with the sickle-cell trait have from 20 to 40 percent Hb S, and their cells sickle only when blood oxygen tension is greatly reduced (10 mm Hg from a normal of 95 mm Hg) (Stokinger and Scheel, 1973).

The gene for Hb S is most often found in equatorial Africa, parts of India, countries of the Middle East, and countries around the Mediter-ranean. In American blacks of African origin, the incidence of sickle-cell

Glucose-6-Phosphate Dehydrogenase Deficiency and Ozone Toxicity

A genetic deficiency in the red blood cell of the enzyme G-6-PD may lead to hemolytic anemia following exposure to numerous chemical agents. Glucose-6-phosphate dehydrogenase deficiency is a sex-linked human genetic disease. It is principally characterized as a biochemical defect in erythrocytes by which glucose is oxidized in a series of reactions to provide the erythrocyte with reducing power in the form of nicotinamide dinucleotide phosphate (NADPH) and reduced GSH, which are necessary, among other things, to maintain erythrocyte membrane integrity (Beutler et al., 1955; Beutler, 1972). The enzyme G-6-PD catalyzes the first reaction on the pentose monophosphate shunt, and those individuals with G-6-PD deficiency have reduced levels of GSH (Figure 20, p. 103).

Since G-6-PD initiates the only reaction by which NADP is reduced to NADPH, it is suggested that a decrease in NADPH can account for the low level of GSH in G-6-PD–deficient individuals, because NADPH is the necessary coenzyme for the reduction of GSSG to GSH (Beutler, 1972; Paniker, 1975). Under hemolytic stress conditions, deficients are unable to maintain proper erythrocyte membrane integrity.

Individuals with the G-6-PD deficiency were first recognized as medically important when they exhibited extreme hemolytic sensitivity of erythrocytes on administration of antimalarial drugs, including sulfonamides, nitrofurans, analgesics, sulfones, and others (Tarlov et al., 1962; Beutler, 1959). Numerous chemicals encountered by industrial workers have very similar chemical structure and toxicologic activities as those of the hemolytic antimalarial drugs. Theoretically, any chemical compound or its metabolite which can accept hydrogen in the defective red blood cell may be suspected as a possible hemolytic agent (Stokinger and Scheel, 1973; Jensen, 1962; Stokinger and Mountain, 1963). A partial list of such chemicals has been reported and is shown in Table 8.

In addition to the medicinal and industrial chemicals previously mentioned as agents in precipitating hemolysis in G-6-PD–deficient individuals, other factors, including additional environmental exposures, preexisting organic disease, and viral and bacterial infections, may be involved. The interaction of these factors may be additive or even synergistic (Stokinger and Mountain, 1963). For example, increased susceptibility to hemolysis is enhanced by hyperthyroidism, since it lowers blood GSH (Lazarow, 1954; Jocelyn, 1958). Furthermore,

Table 8. Compounds Known to Induce Hemolysis of Primaquine-Sensitive (G-6-PD–Deficient) Red Blood Cells

Primaquine	Sulfacetamide
Pamaquine	Thiazolsulfone
Pentaquine	Antipyrine
SN 3883	Probenecid
CN 1110	Nitrofurantoin
SN 15324	Acetylsalicylic acid
Sulfanilamide	Furazolidone
Acetanilid	Sulfamethoxypyridazine
Phenylhydrazine	Salicylazosulfapyridine
Sulfoxone	Naphthalene
Acetophenetidin	Para-aminosalicylic acid
N_2 Acetylsulfanilamide	Pyramidone

Source. Stokinger and Mountain (1963). Test for hypersusceptibility to hemolytic chemicals. *Arch. Environ. Health* **6:** 57. Copyright 1963, American Medical Association.

animal studies have indicated that hyperthyroidism enhanced susceptibility to the toxic effects of carbon tetrachloride as well as lead, a substance having direct effects on red blood cell GSH levels (Kasbekar et al., 1959). Other hormones influence red blood cell GSH levels (Jocelyn, 1958).

The remainder of this section on G-6-PD–deficient individuals and ozone toxicity is based on the theoretical study by Calabrese et al. (1977),[*] which has indicated a possible relationship of ozone exposure with an acute hemolytic response in G-6-PD–deficient individuals. Included in the subsequent narrative is the development of a dose-response relationship of such a phenomenon as well as the theoretical foundations on which it is based.

Buckley et al. (1975) have reported that inhaled ozone causes several physical and biochemical changes (e.g., increased lytic sensitivity and decreased GSH levels) affecting the membrane stability of red blood cells of normal human individuals.

Individuals with a G-6-PD enzyme deficiency have been reported to have significantly lower levels of reduced GSH as compared with "normal" individuals. The levels of GSH in whole blood of normal white adults range from 53 to 84 mg percent (Zinkham et al., 1958). In con-

[*] With the permission of the *Journal of Toxicology and Environmental Health* and the Hemisphere Publishing Corporation, 1977.

trast, individuals with G-6-PD deficiency have whole blood GSH levels from 38 to 51 mg percent (Beutler et al., 1955).

Glutathione is a tripeptide of cysteine, glutamic acid, and glycine which constitutes over 95 percent of the reduced nonprotein sulfhydryl compounds in the erythrocyte (Beutler et al., 1955). The presence of GSH is necessary to maintain the stability of sulfhydryl-containing enzymes. Reduced GSH is also bound to at least one enzyme, is a cofactor for another, and protects hemoglobin, several enzymes, and coenzymes from oxidation (Tarlov et al., 1962).

Statistical associations based on animal studies have been used to establish a relationship between the rate of spontaneous hemolysis and the degree of GSH oxidation. Felger (1952), using horse blood, indicated that there is no marked increase in the rate of hemolysis until the GSH levels decrease to approximately 40 percent of its mean initial value.

In addition to these animal studies, human data reveal that a reduction in the level of GSH in whole blood of approximately 34 to 60 percent below normal (i.e., in individuals with no G-6-PD deficiency) is frequently associated with the precipitation of acute hemolytic anemia (Zinkham et al., 1958). The basis of the human data is summarized as follows:

1. G-6-PD–deficient individuals usually have only 60 to 70 percent of the total GSH of normal individuals when both groups are not under "oxidant" drug stress (Beutler et al., 1955).
2. Acute hemolytic anemia in G-6-PD–deficient individuals treated with primaquine (30 mg) daily is associated with a 14 to 20 percent reduction in GSH levels (Tarlov et al., 1962).

It should be made clear that we are not saying that the reduction in the levels of reduced GSH is the cause of hemolysis. Other mechanisms for red blood cell destruction resulting from oxidant stress have been suggested, including the action of hydrogen peroxide (H_2O_2), lipid peroxidation, the toxic effects of GSSG, binding of Heinz bodies to the membrane, and others (Beutler, 1972). Reduced GSH levels are being used here as an indicator of membrane stability and hemolysis onset, and not necessarily the causal agent.

Biochemical research has shown that hemolytic agents produce H_2O_2 in intact red blood cells and that the oxidative effects seen in hemolysis (e.g., loss of GSH) are indications of the presence of H_2O_2 (Cohen and Hochstein, 1964). Further studies have demonstrated that atmospheric ozone causes reduced GSH to be irreversibly oxidized (Menzel, 1971) and that ozone also produces H_2O_2 in the whole blood (Goldstein, 1973).

Incubation of erythrocytes with ozone likewise causes hemolytic responses in the red blood cells (Goldstein and Balchum, 1967).

Previously cited studies by Buckley et al. (1975) concerning the effects of ozone on "normal" males have established that a 0.5 ppm exposure for $2\frac{3}{4}$ hr effects a 14 percent decrease in GSH values. Concomitant with reduced GSH levels, these individuals have demonstrated homeostatic compensatory adaptation by increasing the activity of G-6-PD by 20 percent. Such an adaptation assists in stabilizing GSH levels. These adaptive "stabilizing" abilities, so evident in normal individuals, are notably diminished in the G-6-PD–deficient individuals.

Thus, if one attempts to determine the percentage of reduction in GSH which would have resulted if a G-6-PD–deficient individual had been exposed to the identical amount of ozone, several assumptions are required:

1. The mechanism of action of the drugs used in the glutathione "stability" test affects the GSH levels in an identical fashion as ozone.
2. There is a dose-response relationship between ozone exposure and a hemolytic response.

After exposure to oxidant drugs during the GSH stability test (Beutler, 1957), levels of GSH in normal individuals drop by approximately 20 percent, at which time the levels of GSH stabilize (Zinkham et al., 1958). However, under identical conditions the GSH level of a G-6-PD–deficient individual does not stabilize and often decreases more than 80 percent and may even reach 100 percent depletion (0.0 mg percent). Clinical tests have indicated that whole blood from patients with drug-induced hemolytic anemias show GSH levels equal to or less than 20 mg percent (Zinkham et al., 1958). The principal reason for this diminished capacity to stabilize GSH levels is the deficiency of the G-6-PD enzyme.

As the result of these data and the above-mentioned assumptions, it is calculated that a 14 percent decrease in the GSH levels in "normals," as occurred after 0.5-ppm ozone (or 0.4-ppm according to Los Angeles County analytical methods) exposure for $2\frac{3}{4}$ hr is approximately equal to a 56 percent reduction in G-6-PD–deficient individuals. Consequently, a G-6-PD–deficient individual with a typical GSH level of 40 mg percent and exposed to 0.5 ppm ozone for the $2\frac{3}{4}$-hr time period could, based on these calculations and assumptions, show a GSH level of approximately 18 mg percent. Levels of GSH equal to or below 20 mg percent in the

GSH stability test are usually present in G-6-PD–deficient individuals and are associated with the onset of acute hemolysis.

Thus, it is predicted that individuals with a G-6-PD deficiency should be considered at high risk to the hemolytic action of breathable ozone. Furthermore, the model developed here predicts that a G-6-PD–deficient individual may experience an acute hemolytic crisis following less than 3 hr of ozone exposure at 0.5 ppm.

More than 80 variants of G-6-PD have been identified by a variety of analytical techniques (Beutler, 1972). The most common variant found in the United States is the A-type. Individuals with the A-type of variant are susceptible to hemolysis on exposure to exogenous hemolytic agents such as drugs (e.g., primaquine) and are expected to be at high risk to ozone.

The incidence of this trait is very high among United States blacks (approximately 11 percent of black males) (Beutler, 1972). The incidence of primaquine sensitivity (i.e., G-6-PD deficiency in male subjects) includes Caucasians: Americans, 0.1 percent; British, 0.1 percent; Greeks, 1 to 2 percent; Sardinians, 1 to 8 percent; Indians from India, 0.3 percent; Mediterranean Jews, 11 percent; and European Jews, 1 percent; and Mongolian: Chinese, 2 to 5 percent; and Filipinos, 12 to 13 percent (Stokinger and Mountain, 1963; Lazarow, 1954). Considerable concern should be directed to G-6-PD deficients who live in "high ozone" cities such as Los Angeles, Philadelphia, and Chicago, where respective levels of 0.6, 0.3 (Altshuller, 1975), and 0.2 ppm ozone (Masterson, 1975) may occur from April to October.

Deficiencies of Catalase and Toxicity to Oxidizing Agents

Another disease which can affect the function of red blood cells is acatalasemia, which is a human genetic disorder inherited as an autosomal recessive (Aebi and Suter, 1972). It is characterized as a deficiency in catalase, an enzyme responsible for the breakdown of H_2O_2. Acatalasemia refers to the homozygous condition, and hypocatalasemia denotes the heterozygous carrier. Homozygotes for the trait generally have little or no catalase activity, and heterozygotes have approximately half the normal level (Takahara et al., 1960; Takahara, 1967). Catalase is normally found in all tissues and is especially high in the liver, kidneys, and erythrocytes.

The acatalasemic condition is most often found without clinical symptoms. The erratic occurrence of symptoms preferentially affects children and usually begins as a small painful oral ulceration in crevices around the neck of teeth, which is thought to be the result of H_2O_2-producing bacteria present in the mouth (Aebi and Suter, 1972). The symptoms can appear in mild, moderate, and severe forms.

The physiological role of catalase remains unclear at present. In erythrocytes, it is believed that catalase is important at only high levels of H_2O_2 stress and GSH-P_x is responsible for the elimination of low levels of H_2O_2 in the blood (Cohen and Hochstein, 1963; Jacob et al., 1965). Under physiological conditions, the ability of the erythrocyte to protect hemoglobin from oxidation by H_2O_2 depends on the presence or absence of glucose, with GSH-P_x critically important in the scheme (Figure 20, p. 103). Erythrocytes, which are either acatalasemic or treated in such a way as to inactivate catalase, are protected from oxidative hemolysis by H_2O_2 by a compensatory stimulation of the HMP shunt, which results in an increased activity of the GSH regeneration mechanism (Jacob et al., 1965). Acatalasemic erythrocytes under nonstress conditions metabolize glucose through the shunt at 3 times the normal rate. Under H_2O_2 stress conditions, glucose is metabolized at 12 times the normal rate in the shunt. This provides the underlying mechanism for oxidative protection.

Previously, it was shown that individuals with the G-6-PD deficiency are sensitive to oxidative stress (Beutler, 1972). Glucose-6-phosphate dehydrogenase catalyzes the first reaction in the HMP shunt, and deficiencies in this enzyme result in decreased levels of reduced GSH, a necessary component for GSH-P_x to function in breaking down H_2O_2. From the foregoing, it might be predicted that erythrocytes devoid of catalase and also deficient in HMP shunt activity would be extremely vulnerable to oxidant damage by peroxides. When erythrocytes of a G-6-PD–deficient individual were inhibited for catalase activity, the presence of H_2O_2 caused a rapid accumulation of irreversibly denatured hemoglobin (Jacob et al., 1965). An individual who had inherited both G-6-PD deficiency and acatalasemia was indeed susceptible to the development of severe hemolytic anemia (Szeinberg et al., 1963). The relationship of G-6-PD deficiency and catalase deficiency is not yet understood. In a screening of 200 American blacks, decreased catalase activity was present only in association with the G-6-PD deficiency (Tarlov and Kellermeyer, 1961). The catalase activity of the G-6-PD deficients was only 60 percent of normal blacks.

Catalase deficiency was first recognized in Japan and Korea and has now been detected in Western countries (Switzerland, Israel, and Germany). Observations from the United States suggest the presence of heterozygotes. No gene frequency data exist for the United States, but, in Japan, the gene frequency is approximately 0.0025, and with a total population of 212 million (approximate United States population), this would amount to approximately 3800 acatalasemics and 1,060,000 hypocatalasemics (Aebi and Suter, 1972). Using the Swiss gene frequency of 0.012, these values would amount to 16,128 acatalasemics and 5,026,994 hypocatalasemics in a population of 212 million (Aebi and Suter, 1972). It should be noted that the frequency of acatalasemia and hypocatalasemia in the United States would probably be different from the calculated frequencies in both Japan and Switzerland. The calculations made here only serve to provide an extremely rough approximation of their incidence in a population number similar to that present now in the United States.

Radiation Sensitivity as Affected by Genetic Factors

Taylor et al. (1933) first suggested that catalase might be able to modify the effects of ionizing radiation. They noted that since ionizing radiation is known to affect the production of peroxides such as H_2O_2 in tissue and since catalase is known to detoxify H_2O_2, it is reasonable to hypothesize that catalase may have a modifying effect on the action of radiation on the body. Experiments by Evans (1947) supported the hypothesis of Taylor et al. when he demonstrated that crude extracts of sea urchin sperm containing catalase could prevent the death of sperm immersed in irradiated seawater. Barron and Dickman (1949) showed that the addition of catalase (0.3 μg/ml) partially prevented the inactivation of crystalline phosphoglyceraldehyde dehydrogenase by both a- and β-radiation. They indicated that H_2O_2 contributed about 30 percent of the total inhibition caused by a-radiation and somewhat more for inhibition by β-rays.

A Swiss study using human subjects found that red blood cells from acatalasemic people, when exposed to irradiation, form MetHb about 10 to 20 times faster than "normals" (Aebi et al., 1962). Furthermore, the effect of γ-irradiation on cancer cells is tremendously reduced by simply adding catalase to the outside of the cell (Thomson, 1963).

Catalase has been found to be present in both the nucleus and cytoplasm. However, it seems that about 90 percent of the catalase is apparently present in the cytoplasm (Thomson, 1963; Ludewig and Chanutin, 1950; Creasy, 1960). The location of catalase in the cell is considered important with respect to the metabolism of its substrate (H_2O_2) (Thomson, 1963). Consequently, cytoplasmic catalase may protect against the formation of MetHb or changes in cell membranes, and nuclear catalase is thought to offer limited protection to the DNA from H_2O_2. Since only a small percentage of catalase is present in the nucleus, it seems that catalase may not be highly effective in protecting the genetic material from oxidant stress. However, since the volume of the nucleus compared with the entire volume of the cell is often quite small, there may not be the low levels of catalase in the nucleus as compared with levels in the cytoplasm if measured on a per volume basis.

It has been found that catalase actually plays a secondary role to GSH-P$_x$ in the detoxification of H_2O_2, and, in fact, only when the concentration of H_2O_2 exceeds a certain undefined high level does catalase serve a significant role in H_2O_2 detoxification. This finding may help to explain the widespread conflicting reports concerning the protective effects of catalase on the action of radioactivity. For example, in contrast to previous reports indicating the importance of catalase with regard to detoxifying H_2O_2 following irradiation, a study with hypocatalasemic guinea pigs exposed to various levels of irradiation showed they were not significantly different from the controls with regard to radiation sensitivity and longevity (Rusev et al., 1960).

The dose of radiation which produces the "critical" concentration is probably modified by several factors, including natural levels of catalase in the various tissues (levels are specific for each tissue) and serum vitamin E (the absence of vitamin E is associated with the accumulation of peroxides), vitamin C, and selenium (Se) levels.

However, the principal way in which the cell detoxifies H_2O_2 (reduces oxidant stress) is by the action of the enzyme GSH-P$_x$. Figure 20 shows the relationship of GSH-P$_x$ in the normal biochemical scheme. Thus, in order to maintain continuous functioning of the enzyme, it is necessary to form reduced GSH continuously, which in turn is dependent on other factors such as the supply of glucose, the enzyme G-6-PD to reduce NADP to NADPH, and GSSG-reductase.

In support of this scheme, human studies, *in vitro*, on the effects of peroxide on normal cells, acatalasemic cells, and cells deficient in G-6-PD indicated that only cells deficient in the G-6-PD showed marked decreases in reduced GSH levels (Powers, 1963). It should be emphasized that such changes are closely associated with increased cell membrane fragility (see section on vitamin C and radioactivity, p. 101).

Methemoglobinemia and Its Causes

Methemoglobin is formed when the ferrous ion of the hemoglobin molecule is oxidized to its ferric state. While in the ferric state, hemoglobin is unable to bind oxygen reversibly (Keity, 1972). Thus, individuals with methemoglobinemia have a reduced oxygen-carrying capacity of the blood, as well as a decreased capacity of the residual oxyhemoglobin to dissociate and release oxygen to the tissues such as the brain and muscles, which are quite sensitive to reduced oxygen tension. The chemical effects of the accumulation of MetHb are directly related to its quantity in the blood. Many consider levels of 1 to 2 percent MetHb (Gruener and Shuval, 1970; Bodansky, 1951; Committee on Nitrate Accumulation, 1972) as normal in the blood of adults. Usually at levels of less than 5 percent, there are no external signs. At levels of 5 to 10 percent, clinical symptoms such as cyanosis may appear. Death results at levels of 50 to 75 percent (Committee on Nitrate Accumulation, 1972; Knotek and Schmidt, 1964; EHRC-Nitrates, 1974).

Several factors which influence the formation of MetHb include genetic effects, developmental effects, lack of gastric acidity, and dietary (environmental) factors.

Genetic Factors

Most of the reduction of MetHb is accomplished by the enzyme NADH-MetHb reductase (Keity, 1972; Jaffe, 1966). A genetically inherited disorder of NADH-MetHb reductase deficiency has been reported to occur among Eskimos and certain Athapascan groups (a linguistic family of North American Indians ranging from Alaska to northern Mexico and including the Navajo and Apache tribes) (Shearer et al., 1971). Deficiencies in GSSG-R, GSH-P_x, and G-6-PD also predispose individuals to elevated MetHb levels (Beutler, 1972). Although most of these disorders are rare, G-6-PD deficiency is fairly common in American black males (11 percent) and in certain Mediterranean groups (see p. 49).

Developmental Factors

The higher levels of MetHb of 2 to 5 percent in the blood of infants under 6 months of age is thought to be due to the fact that fetal hemoglobin is easily oxidized, as well as their temporary deficiency in MetHb reductase or its coenzyme (Lee, 1970). Further, newborns lack gastric acidity,

which allows nitrate-reducing organisms to thrive in the upper gastrointestinal tract, where nitrates are reduced to nitrites before the
latter can be completely absorbed (Cornblath and Hartman, 1948). The
nitrite ion converts hemoglobin to MetHb.

The lack of sufficient gastric acidity (achlorhydria: absence of free
hydrochloric acid in the stomach) due to diseases such as gastric carcinoma, gastric ulcer, pernicious anemia, adrenal insufficiency, or
chronic gastritis may permit colonization of the upper gastrointestinal
tract by bacterial flora which are able to convert nitrates to nitrites.
Individuals with achlorhydria may, therefore, suffer the consequences
of nitrite-induced methemoglobinemia (EHRC-Nitrates, 1974).

Dietary (Environmental) Factors

The nitrite ion converts hemoglobin to MetHb by oxidizing the ferrous
ion to ferric (Lee, 1970). Human exposures to nitrates and nitrites in
both food and water are now regulated by the federal government,
since high exposure is known to cause increased MetHb levels in the
blood.

The maximum amounts of 200 ppm of sodium nitrite and 500 ppm of
sodium nitrate in food additives were established under the federal
Meat Inspection Act (Anonymous, 1971). Ingestion of various foods with
high levels of nitrates has been reported to cause high levels of MetHb in
infants (Simon, 1966; Commoner, 1970; Keating et al., 1973). High
concentrations of nitrates are often found in spinach, beets, cauliflower,
lettuce, celery, radishes, kale, and mustard. Analysis of commercial
baby foods in Canada (spinach, beets, squash) has nitrate levels with
extraordinarily high values, ranging from nearly 300 to greater than
1300 ppm (Kamm et al., 1965). It should be emphasized that, for nitrates
to present a problem, they must be first converted to nitrites, as previously mentioned.

Another source of nitrates is drinking water. The federal drinking
water standard for nitrate (as N) is 10 mg/l (Public Health Service,
1962). A Russian study has revealed that levels as low as 26 mg/l have
been associated with elevated MetHb levels and slowed motor reflexes
in adolescents (Petuknov and Ivanov, 1970). Nitrate levels in Illinois
have exceeded the federal standards at various times in several rivers.
The highest nitrate concentrations are consistently associated with
areas of intensive agricultural production, in fertile, well-drained soils
naturally rich in organic nitrogen (Harmeson et al., 1971).

Another high-risk group with respect to MetHb formation may be
those with vitamin C deficiency. Vitamin C has been found to reduce

MetHb actively and is used as a therapeutic agent in cases of high levels of MetHb in the blood (Lee, 1970; EHRC-Nitrates, 1974) (see section on vitamin C).

Vanadium is another factor which is known to cause MetHb formation (Mitchell et al., 1968). How much of a potential problem this may be is unknown.

BLOOD SERUM DISORDERS

Serum Alpha$_1$-Antitrypsin Deficiency

Individuals with an inherited deficiency of serum alpha$_1$-antitrypsin (SAT) have been found to be predisposed to alveolar destruction even in the absence of chronic bronchitis. This was first reported by Laurell and Eriksson (1963) and subsequently verified in numerous reports (Tarkoff et al., 1968; Talamo et al., 1973; Talamo et al., 1966; Briscoe et al., 1966; Talamo et al., 1971). Other findings have lent support to the postulate that an inherited deficiency of SAT predisposes an individual to the development of pulmonary emphysema by allowing digestion of human lung tissue by proteases (proteolytic enzymes) from leukocytes or macrophages (Lieberman and Mohamed, 1971; U.S. Dept. HEW, 1971). These proteolytic enzymes include trypsin, chymotrypsin, elastase, collagenase, thrombin, and plasmin (Vogel, 1968; Eisen et al., 1970). Normal human serum is capable of inhibiting these enzymes, thereby protecting the individual from pulmonary emphysema. Research has strongly suggested that the lung is protected from proteolytic enzymes by a series of serum proteins, of which SAT is one. Most of the human serum antitryptic activity (85 to 90 percent) is associated with an alpha$_1$ protein, which is SAT (Jacobsson, 1955). The release of proteolytic enzymes occurs from cells (leukocytes and macrophages) that are responsible for cleaning the surfaces of the alveoli and bronchioles and removing foreign material and bacteria from these regions (Lieberman, 1972; Janoff, 1972; Cohn and Wiener, 1964). A wide variety of different challenges, both particulate and gaseous, and particularly both of these together, leads to an increase of these two types of cells in the lung as a more or less inevitable consequence either of direct irritation or of deposition in the alveoli of foreign material (Vogel, 1968; Bates, 1972; Kilburn, 1973; Corrin and King, 1970; Wood, 1951; Morrow, 1967; Coffin et al., 1968). A study performed on rabbits exposed to ozone showed an influx of heterophilic leukocytes and a diminution of pulmonary alveolar macrophages obtained by pulmonary lavage. This phenonemon was shown to be dose-related. Additionally, a decrease in the ability of the alveolar macrophages to engulf streptococci was demonstrated. The data suggest that ozone either destroys the alveolar macrophage *in situ* or renders them sensitive to lysis during the process of lavage (Coffin et al., 1968).

A significant number of cases of chronic obstructive lung disease is associated with SAT deficiency in its severe (homozygous) and intermediate (heterozygous) forms (Lieberman, 1969). Blood serum of patients (suffering from pulmonary emphysema) homozygous for the SAT deficiency was found to contain less than 10 percent of the SAT of that of "normal" individuals. It has also been shown that these homozygous individuals (SAT-deficient) tended to develop emphysema in the third or fourth decade of life without symptoms of bronchitis. Additionally, heterozygous individuals were found to have a concentration of SAT between 50 and 60 percent of "normal" (Kueppers et al., 1964). However, not all carriers of the deficiency develop overt lung disease (Mittman et al., 1971).

Although patients with the homozygous condition are quite rare,* nearly 10 percent in certain segments of the population have intermediate levels of SAT (Table 9) (Mittman and Lieberman, 1973). Medical surveys have indicated the intermediate deficiency occurs more frequently than statistically expected in patients with obstructive lung disease, thus implying that the intermediate deficiency also predisposes the development of obstructive lung disease† (Lieberman et al., 1969; Kueppers et al., 1969).

Some individuals with severe SAT deficiency and most with the intermediate deficiency never develop lung disease. However, physiological studies in asymptomatic carriers of SAT deficiency imply that many have the lesions of anatomic emphysema (Mittman et al., 1972). Nonsmoking volunteers with an intermediate deficiency show little evidence of obstructive disease. However, a significantly greater number of apparently healthy individuals with the intermediate deficiency who smoke have early but definite airway obstruction (Mittman et al., 1972). Mittman et al. (1971) have reported that most patients with obstructive lung disease and the intermediate deficiency have smoked and their

---

* An incidence of approximately 0.8 per 1000 was found in a population of 16,190 individuals from Norway and Sweden (Fagerhol and Laurell, 1970).

† It is important to note that emphysema has been observed less commonly in Negroes than whites and appears infrequently in American Indians and Orientals. Variations in the occurrence of SAT deficiency in different ethnic groups might account for these observations (Mittman and Lieberman, 1973).

backgrounds suggest that the disease developed after less exposure to cigarettes than typical patients with normal SAT levels.

A factor which can complicate the interpretation of SAT screening tests is that in certain individuals the levels of leukocyte lysosomal protease-antiprotease are imbalanced (Galdston et al., 1972). They found some individuals with severe SAT deficiency but without the expected clinical phenotype of chronic obstructive pulmonary disease. The finding of low levels of leucolytic protease together with the SAT deficiency in such individuals may offer a possible explanation for the absence of chronic obstructive pulmonary disease in people with severe SAT deficiency.

Additionally, it has been shown that cystic fibrotic (CF) patients generally had normal levels of SAT but that relatively low levels of SAT were found in the sera of many parents of CF children. Particularly, the mean level of sera SAT from the fathers was significantly lower than in the children and the mothers. No final explanation was given for this, but one possibility may be that in CF patients and carriers the secretion of SAT is somewhat hampered. The "normal" mean levels in the children and mothers may have been due to increased levels in patients with active inflammation and the use of oral contraceptives in the mother (Fagerhol, 1973). A theoretical scheme illustrating the inter-

Table 9. Percentage of Individuals of Different Ethnic Background with an Intermediate SAT Deficiency

Group	Percentage of Subjects Affected	Number Tested
Irish	9.0	194
Russian and Central Europe	7.9	63
German	7.0	171
English	6.6	151
French and Belgium	4.7	64
Italian	0	53
Jewish	2.6	114
American Negro	1.4	70
Mexican-American	1.3	76
American Indian	0	37

Source. Mittman and Lieberman (1973). Screening for a_1-antitrypsin deficiency. In Ramot, B., et al. (Eds.), *Genetic Polymorphism and the Diseases in Man.* Academic Press, New York, p. 191. By permission of the author and the Israel Journal of Medical Sciences.

relationships of SAT, G-6-PD, and catalase deficiencies and cystic fibrosis to ozone exposure is shown in Figure 9 (see p. 102 for a discussion of cystic fibrosis and vitamin E absorption).

Chowdhury and Louria (1976) have recently investigated the influence of Cd and other trace metals on human SAT. Cadmium was the only trace metal that reduced the concentration of SAT and depressed the trypsin inhibitory capacity. Divalent Pb, Hg, Ni, and Zn ions showed no such effects. The combination of Cd–Ni, Cd–Pb, and Cd–Fe did not produce greater decreases of SAT activity than did Cd at the same concentration. However, the combination of Cd–Hg appeared to be more effective in decreasing the SAT activity than Cd alone. These results are of significance, since emphysema can be caused by prolonged exposure to Cd (Lane and Campbell, 1954; Holden, 1965). The concentrations of Cd used in the Chowdhury and Louria (1976) study are far greater than concentrations of Cd found in the blood of normal adults (Kubota et al., 1968), but they are similar to blood concentrations in industrially poisoned workers (Friberg et al., 1972). Additionally, in those workers dying with Cd-induced emphysema, concentrations of 50 to 600 mg/g of lung tissue have been noted (Lane and Campbell, 1954; Flick et al., 1971). These were considerably above the concentrations (1 to 5 mg/100 ml) used in the Chowdhury and Louria study. Their study may indicate a possible explanation for the emphysema that occurs in industrial workers exposed to Cd. These results imply that individuals either homozygous or heterozygous for the SAT deficiency may be considered at high risk with regard to both Cd and Cd–Hg exposure.*

Pseudocholinesterase Variants and Insecticide Toxicity

The use of insecticidal agents is an important component of modern farming practice in the United States. One type of insecticidal agent is the chemical insecticide which is designed to disrupt the normal functioning of the insect's nervous system and thus lead to its death.

In order to effect muscle or nerve innervation, the substance acetylcholine (ACH) is first released from the end of a nerve cell. The ACH then migrates to the surface of the adjacent cell in order to effect muscle or nerve innervation. Following the nerve or muscle innervation, ACH is quickly broken down by acetylcholinesterase (ACHase). The action by ACHase is necessary to protect the cell from nerve tremors. Certain insecticides, including the organophosphates and carbamates, are

* See Glaser et al. (1977) for a recent criticism of the Chowdhury and Louria (1976) study.

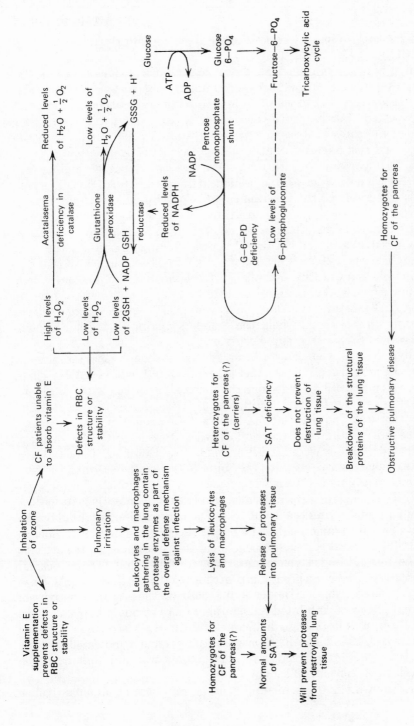

Figure 9. Theoretical considerations of the effects of ozone, serum alpha$_1$–antitrypsin deficiency, G-6-PD deficiency, acatalasemia, and cystic fibrosis on the structure and function of the human lung and RBCs. *Source.* Sorensen (1976). School of Public Health, Univ. of Illinois, Chicago.

designed to inhibit ACHase, thereby leading quickly to the insect's death.

In light of the neurological activity of such insecticides as well as the remarkable similarities between human and insect neurons, the numerous reported cases of human insecticide intoxication are not unexpected. Further investigation of insecticidal action (described below) indicates that segments of the human population may be at increased risk to certain anticholinesterase insecticides.

It is important to note that there are two types of cholinesterase: acetylcholinesterase and pseudocholinesterase (pACHase). As previously mentioned, ACHase inactivates ACH produced at the neuromuscular junction during neurotransmission. Acetylcholinesterase is located in numerous tissues. Pseudocholinesterase, although not present in red cells, is found in most tissues, including plasma. The function of pACHase is still not known; however, it has been suggested that the main role may be the hydrolysis of the certain cholinesters which inhibit ACHase (Lehmann and Liddell, 1972). A summary of normal neuronal structure and function is provided on page 62 for readers needing this review to help their understanding of the subsequent theoretical high-risk group discussion.

Physiological studies have revealed that, although most individuals have the identical pACHase, there are individuals whose pACHase has different metabolic activity from the "normal" or typical type. Biochemical studies indicate that the different pACHase have different molecular structures. The different pACHase types are referred to as variants.

The initial studies of individuals with atypical pACHase variants indicated that they were usually symptomless. However, they exhibited an extreme sensitivity to the muscle relaxant suxamethonium. Suxamethonium has been widely used by anesthetists for the past several decades, since the muscular paralysis conveniently lasts for less than 5 min following the usual dose. The short action is caused by the hydrolysis of suxamethonium by pACHase (Evans et al., 1952; Bourne et al., 1952). However, later biochemical studies comparing the activity of the normal and atypical pACHase revealed that (1) the normal (typical) enzyme has a greater affinity for substrates than the pACHase from sensitive subjects (Davies et al., 1960); (2) the pACHase activity of normal plasma is inhibited considerably more strongly by most of the pACHase inhibitors (e.g., dicubaine and fluoride) (see below) (Kalow and Davies, 1958). Results from such studies clearly indicated that the pACHase present in "sensitive" individuals was different from the pACHase of "normal" plasma.

Research involving screening of large numbers of humans (Kalow and Gunn, 1958; Kattamis et al., 1963) has revealed that the development of atypical or variant types of pACHase is under genetic control. The variants are usually determined by two genetic loci (Ch_1 and Ch_2). The variants at the Ch_1 locus are identified by altered sensitivity of the enzyme to inhibitors such as dicubaine, R02-0683 [the dimethyl carbamate of (2- hydroxy-5-phenylbenzyl) trimethyl ammonium bromide], fluoride, and butanol. An additional rare allele (the silent gene) is also found at the Ch_1 locus; its presence results in either the complete absence (type 1) or slight levels (type 2) of cholinesterase activity (Lehmann and Liddell, 1972).

Gene frequencies have been determined for some of the variant genotypes. The most frequent "atypical" homozygous variant (the dicubaine variant) occurs with a frequency of 1 in 2800 healthy Canadians of European ancestry (Kalow and Gunn, 1958) and has been found to be extremely sensitive to R02-0683 (Lehmann and Liddell, 1972). This is of particular significance in light of the widespread use of carbamate insecticides. It should be pointed out that 3 to 4 percent of the Canadian population tested was found to be heterozygote carriers (normal and dicubaine genes together) of intermediate sensitivity (Kalow and Gunn, 1958). Additionally, of the 10 recognized genotypes of pACHase at the Ch_1 locus, 4 are known to display a marked sensitivity to suxamethonium, specifically, the homozygotes for dicubaine sensitivity and the silent gene and the heterozygotes for the dicubaine–silent gene combination and the dicubaine–fluoride combination. Their combined frequency in individuals of European ancestry is 1 in 1250 (Szeinberg, 1973).

In addition to genetic factors, there are several other conditions in which low pACHase activities have been found, for example, liver diseases (McArdle, 1940), malnutrition(Waterlow, 1950), infectious diseases (Hall and Lucas, 1937), and organophosphorous (OP) poisoning (Barnes and Davies, 1951).

In terms of public health, the data indicate that individuals have differential sensitivity to the activity of various neuromuscular-acting drugs and insecticidelike chemicals. Differences in sensitivity are directly related to the occurrence of cholinesterase enzyme variants and its diminished ability to inactivate the drug or insecticide analog. Individuals with such pACHase variants should be considered potentially at high risk to anticholinesterase insecticides (Lehmann and Ryan, 1956). It should be emphasized that not all drugs and insecticidelike compounds act with greater sensitivity in atypical pACHase variants. For example, two organophosphate insecticides (tetraethylpyrophosphonate (TEPP) and diisopropylfluorophosphonate (DEPisofluorophosphate))do not inhibit differentially (Kalow and Davies, 1958) among pACHase variants and would not cause a higher risk to those with an atypical variant. In a limited study of OP poisonings, Tabershaw and Cooper (1966) noted no increased susceptibility of heterozygous dicubaine carriers.

Neuronal Structure and Function*

The central nervous system (CNS) in mammals includes the brain and spinal cord; in insects, the CNS refers to the chain of ventral ganglia which may be partially fused together in some insects (e.g., the housefly) or arranged as discrete segmented sections in other insects (e.g., the cockroach). The CNS is that part of the nervous system which integrates the broad diversity of information received from the environment. The peripheral nervous system is composed of the afferent nerves, which provide data to the CNS, and the efferent nerves which take "orders" to muscles, glands, and so on. The functional unit of the nervous system is the neuron, or nerve cell. Neurons are cells with long processes along which nerve impulses are conducted. The "typical" neuron has one long process called the axon, which conveys the nerve impulse from the cell body to the short branched processes (the dendrites) which carry impulses to the cell body of the next neuron (Figure 10).

In order to conduct a nerve impulse, it is necessary that two different types of transmission be functional. The *axonic transmission* transports

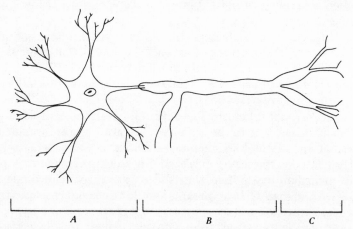

A   B   C

Figure 10. Conceptual diagram showing the three functional regions of nerve cells: A, generator region; B, conductile region (axon); C, transmissional region (dendrites).

* See general references such as O'Brien (1967) and Ochs (1971, 1971a).

an impulse from its entry into the nerve cell and then along the axon to the dendrite, which is in extemely close proximity to the cell body of a subsequent neuron, muscle, or gland. Across the space between the two cells, *synaptic transmission* takes place. The term "synapse" refers to the junction of neurons or to the junction between neuron and muscle.

With regard to the internal electric potential of the axon, the inside environment is more negative than the outside. Such an axon is said to be polarized. The *resting potential* difference between the inside and outside of the axon is called the *membrane potential*. As a nerve impulse passes along the axon, the inside immediately becomes more positive than the outside. However, the original preimpulse condition is quickly reestablished as the impulse passes on. The moving depolarization is called an *action potential*.

The resting potential is caused by the presence of a higher Na^+ level outside the membrane than inside. The action potential is caused by the influx of Na^+ through the outer membrane of the axon. Thus, as the Na^+ enters the cell, the charge on the outside quickly drops to zero and then even goes positive. As the membrane becomes permeable to K^+—which, since the charge is greater on the inside than outside, passes out—the original electrical difference is reestablished. Sodium ions are then "pumped" out of the axon, and K^+ back in, in order to maintain the original low level of Na^+ in the axon and thus its ability to be fired.

When the impulse reaches the synapse, it ends or dies. However, it effects the release of a neurotransmitter from a synaptic vesicle located in the dendritic part of the cell. This substance is known to diffuse across the synapse and effect the initiation of another action potential if the synapse is between neurons, or an appropriate response if the synapse is between a neuron and some effector (e.g., muscle or gland). There are two broadly accepted neurotransmitters: acetylcholine and nor-epinephrine (Figure 11).

Synapses which have acetylcholine are called cholinergic and those with norepinephrine are referred to as adrenergic. It is the function of the neurotransmitter to combine with the receptor on the postsynaptic side to cause a structural change of the membrane which affects the Na^+ permeability, thereby initiating an action potential.

To restore the receptor to its resting potential, the transmitter must be eliminated. At cholinergic junctions, it is quickly removed by cholinesterase (ACHase). ACHase hydrolyzes ACH to acetate and choline. The ACHase is usually located on the presynaptic side of the synapse. Since the synaptic gap is only approximately 500 Å wide and each ACHase molecule is about 9 Å long, having ACHase located on the presynaptic side should not impede rapid hydrolysis of ACH by ACHase.

In adrenergic junctions, the equivalent enzyme is monoamine oxidase.

Peripheral nerves are either somatic or autonomic nerves. Sensory somatic nerves are afferent, and motor nerves, carrying impulses to the voluntary muscles, are efferent. The autonomic nerves are all efferent and convey nerve impulses to glands and involuntary muscles (e.g., intestinal muscles). The autonomic nervous system is made up of two components: the sympathetic and the parasympathetic.

$$CH_3C(O)OCH_2CH_2{}^+N\ (CH_3)_3$$

Acetylcholine

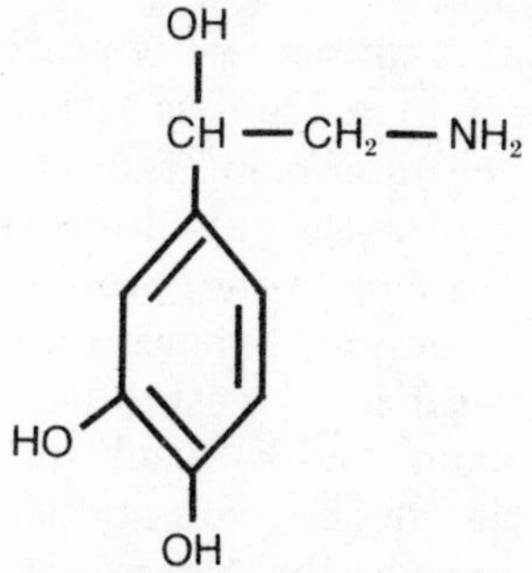

Norepinephrine

Figure 11. The chemical representation of the two most broadly recognized neurotransmitters.

Axons and synapses are present in both the CNS and the peripheral nervous system. Somatic and autonomic nerves have synapses in the CNS. However, the only peripheral synapse of somatic nerves is the neuromuscular junction. In autonomic nerves, there are always two peripheral synapses: one at the junction of the muscle or gland and another immediate one.

The peripheral synapses of the sympathetic nerves are adrenergic, except for the adrenal medulla and the sweat glands. The peripheral synapses of the parasympathetic nerves are all cholinergic. However, the intermediate synapses of both parasympathetic and sympathetic nerves are cholinergic.

HOMEOSTATIC—REGULATORY DISORDERS

Genetic Diseases (Cystinuria, Cystinosis, and Tyrosinemia) of the Kidney and Heavy Metal Toxicity

A broad variety of factors, including genetic diseases, nutritional deficiencies, and environmental pollutants, may adversely affect the functioning of the kidney so that in certain cases abnormally high urinary levels of water, phosphate, sodium, potassium, glucose, amino acids, and other organic acids may occur (Schneider and Seegmiller, 1972).

Several of the more important genetic diseases which may adversely affect resorption of nutrients by the kidney, resulting in Fanconi's syndrome (hyperaminoaciduria, glycosuria, and hypophosphatemia) include (1) cystinuria, (2) cystinosis, and (3) tyrosinemia (Schneider and Seegmiller, 1972). For individuals desiring a brief summary of normal kidney structure and function, turn to page 67.

Genetic

Cystinuria. Cystinuria is a disorder of amino acid transport affecting the epithelial cells of the renal tubule. The disease is usually noted by the precipitation of cystine in the urethra. Other amino acids, including lysine, arginine, ornithine, and cysteine-homocysteine mixed disulfide, are also present in excess in the urine. Cystinuria is usually of no clinical importance except under conditions of low dietary protein and when cystine crystals may develop in the urethra. It has been estimated that about 1 per 200 to 250 individuals excretes excess cystine in the urine and 1 per 20,000 to 100,000 is homozygous for the trait (Thier and Segal, 1972).

Cystinosis. Cystinosis is noted by a high intracellular content of cystine located in particular cell types. Symptoms of the most severe form, nephropathic cystinosis, result in the above-cited substances present in excess in the urine. The continued loss of excessive quantities of phosphate results in the development of hypophosphatemic rickets, resistant to treatment by vitamin D. Growth is often severely retarded, and death usually occurs by the age of ten from uremia. In less severe nephropathic cases (intermediate forms), clinical symptoms are considerably reduced, and such people live to the second or third decade of life. In the least severe form (benign cystinosis), the principal clinical difference between the benign and the nephropathic form of the disease is the absence in these patients of renal dysfunction. However, they do exhibit crystalline deposits in the cornea, bone marrow, and leukocytes, but these do not result in disability, and the patient lives to adult life. Both the nephropathic and benign types of cystinosis have an autosomal recessive form of inheritance. The frequency of the disease in the population has not been assessed (Schneider and Seegmiller, 1972).

Tyrosinemia. In individuals with tyrosinemia, p-hydroxyphenyllactic acid accumulates and is thought to be the cause of tubular damage in such individuals (Gentz et al., 1965). It is of considerable interest that these same compounds accumulate in individuals with scurvy, since ascorbic acid is needed by the enzyme (p-hydroxyphenyl-pyruvic acid oxidase), which is deficient in tyrosinemia. The clinical description is highly variable, depending on whether the disease is chronic or acute. However, very often, afflicted individuals exhibit congenital cirrhosis of the liver, renal tubular defects with aminoaciduria, and vitamin D–resistant rickets (hypophosphatemic condition). Although the frequency of tyrosinemia in the general population is not known, it has been estimated that the frequency of heterozygote carriers in the Chicoutimi region of north-eastern Quebec is 1 carrier for every 20 to 31 people (Laberge and Dallaire, 1967; LaDu and Gjessing, 1972).

Nutritional

Jonxis and Wadman (1950) reported an increased output of some free and bound amino acids in the urine of a child with scurvy. Other clinical studies with two children with scurvy also revealed an increased output of amino acids. However, following ascorbic acid administration for at least 3 weeks, normal levels of amino acids in the urine were approached (Jonxis and Huisman, 1954). Dinning (1953) has reported that vitamin E deficiency in rabbits also acts to increase the levels of amino acids in the urine significantly.

Environmental Pollutants

Hyperaminoaciduria has been reported in experimental plumbism with clinical lead intoxication (Wilson et al., 1953; Bickel and Souchon, 1955; Maggioni et al., 1960) and in asymptomatic industrial workers exposed to Pb (Clarkson and Kench, 1956). Chisolm (1962) further showed Pb intoxication may also produce glucosuria and hypophosphatemia as well as hyperaminoaciduria. According to Chisolm (1962), Pb-induced aminoaciduria is similar to that found in vitamin D–deficiency rickets, cystinosis, hyperparathyroidism with renal calcinosis, and congenital renal aminoaciduria. These conditions result in a generalized renal aminoaciduria in which the reabsorption of all amino acids by the proximal renal tubular cells is uniformly impaired. Chisolm (1962) suggested that "some renal tubular mechanism which is common to and essential for the reabsorption of all amino acids is affected" in these diseases (see page 69 for a contrary view).

Clarkson and Kench (1956) have demonstrated that aminoaciduria may occur in asymptomatic industrial workers exposed to Cd, Hg, and uranium (U) in addition to Pb. Exposure to Cd and U resulted in more extreme changes than did Pb and Hg in the urinary amino-nitrogen levels and the amino acid patterns. Other chemical agents, including maleic acid (Harrison and Harrison, 1954) and Cu in Wilson's disease (Bearn et al., 1957; Stein et al., 1954), have also produced the Fanconi syndrome.

Renal tubular impairment has also been caused by intoxication from a variety of aromatic compounds. Otten and Vis (1968) have noted that ingestion of methyl-3-chromone induced the Fanconi syndrome. They also pointed out the structural similarity of methyl-3-chromone to tetracycline. Benitz and Diermeier (1964) had previously indicated that anhydro-4-epitetracycline, which is the degradation product of tetracycline, is toxic to the renal tubules.

If the suggestion by Chisolm (1962) that the same renal tubular mechanism which is common to and essential for the resorption of all the amino acids is similarly affected by the various genetic diseases, nutritional deficiencies, and pollutants, such as Pb and Cd, be selected as a working model, individuals affected by one or more of the related genetic diseases or nutritional deficiencies should be considered at high risk to the effects of Pb, Cd, Hg, and U. Since these individuals are already predisposed to the development of the Fanconi syndrome, the concomitant exposure of such pollutants would provide an additional stress at a critical biological weak point. It is not possible at this time to determine whether the exposure of these substances, separately or together, in these instances (e.g., genetic disease or nutritional deficiency) would result in toxic effects of an additive or synergistic nature.

Structure and Function of the Kidney

In man, the kidneys are bean-shaped structures about 4 in. long, situated behind the body cavity on either side of the middorsal line. The kidneys are found outside of the body cavity, and only one surface of each kidney is covered with peritoneum. Renal arteries and veins are supplied to the kidney. A tube called the ureter leads from the kidney and empties into the urinary bladder. Another tube, the urethra, transports the urine from the bladder to the outside of the body. A longitudinal section of the kidney reveals an outer layer, the cortex, and an inner layer, the medulla.

The human kidney is made up primarily of many small tubules. In the cortical region, these tubules are coiled, but in the medulla they are straight. A "urinary unit" is made up of one renal or malpighian corpuscle and its related uriniferous tubule. There are about 1 million such units in each human kidney. The renal corpuscle consists of an outer, thin, double-walled capsule, renal (Bowman's) capsule, and this surrounds a dense grouping of blood capillaries called a glomerulus. The capsule extends from the renal corpuscle or the uriniferous tubule. An afferent arteriole passes into every glomerulus, and an efferent arteriole departs from each one. Each coiled uriniferous tubule leads into a larger collecting tubule that extends through the medulla and empties into the renal pelvis. Each uriniferous tubule is modified into three regions: the proximal convoluted tubule, the loop of Henle, and the distal convoluted tubule. The efferent vessel that departs from each glomerulus develops a capillary net about every uriniferous tubule. These capillaries become small veins which join with the renal veins.

The formation of urine is dependent on three separate but related processes: filtration, resorption, and tubular secretion. When the blood goes through the glomerulus, it is under pressure, since the afferent vessel passing into a glomerulus is actually an artery and the efferent vessel is smaller than the afferent one. As a result of the pressure, the liquid part of the blood is filtered and travels to the cavity of the glomerulus. The liquid filtered from the blood is the same as the blood plasma without the blood proteins and other colloidal substances. However, by the time the liquid reaches the end of the uriniferous tubule, it is the composition of urine. This happens primarily because the epithelial cells lining the tubules resorb certain substances from the liquid and also by passing certain substances into the lumen of the tubule (Selkurt, 1971).

During a typical day, the kidney forms between 75 and 150 liters of the filtered liquid, even though the normal amount of urine excreted each day is about 1 to $1\frac{1}{2}$ liters. Thus, large quantities of water are resorbed each day via the tubules. The amount of resorption of other substances varies. There is a specific type of selective resorption by the tubules; for example, urea, uric acid, and phosphates are resorbed to a small degree, but creatinine is not resorbed at all. This resorption occurs primarily in the proximal convoluted tubules. Another mechanism in the determination of urine composition is the tubular secretion factor. In this instance, substances are secreted directly from blood in the renal portal vein. With regard to amino acids, glycine, arginine, lysine, proline, and hydroxyproline are relatively poorly reabsorbed, but others such as histidine, methionine, leucine, isoleucine, tryptophan, valine, threonine, and phenylalanine are effectively reabsorbed. There are probably three renal tubular mechanisms for reabsorption of amino acids: one transports lysine, arginine, ornithine, cystine, and histidine; a second involves aspartic and glutamic acid; a third handles proline, hydroxyproline and glycine (Selkurt, 1971).

Thus, in summary, tubular resorption of different substances is highly variable, depending on the specific substance. In certain instances, for example, the sugars and certain amino acids, a common transport mechanism may be used. However, there seem to be numerous other transport mechanisms for other substances which act independently (Selkurt, 1971).

Porphyrias and Lead Exposure

Metabolic disorders associated with the synthesis of porphyrins, usually in erythroid and liver cells, are collectively known as porphyrias. This group of closely related diseases is characterized by a marked increase in the synthesis and excretion of porphyrins or their precursors. A porphyrin is any one of a group of iron-free or magnesium-free cyclic tetrapyrrole derivatives. Porphyrins are vitally important molecules, some of which serve as intermediates in the synthesis of hemoglobin, the cytochromes, chlorophyll, and vitamin B_{12}. The porphyrin ring is shown in Figure 12. The scheme of heme biosynthesis is shown in Figure 13.

The excessive production of porphyrins leads to abnormally high levels of such substances in the urine and various tissues of the body. Porphyrins often have a photosensitizing activity which produces skin lesions, prickling, itching, and burning following exposure to light (Marver and Schmid, 1972). It is thought that these skin manifestations are influenced by chemicals produced from photochemical reactions which are initiated by the light energy absorbed by porphyrins. Rimington et al. (1967) have suggested that the release of lysosomal enzymes plays a role in the inflammation process.

Besides their photosensitizing properties, porphyrins produce a broad

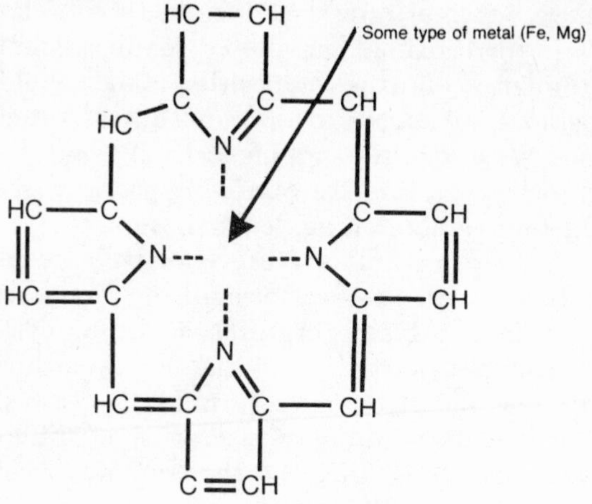

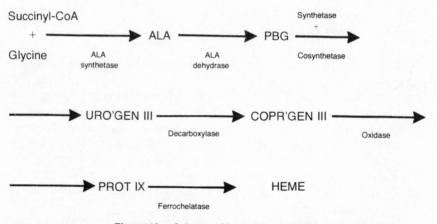

Figure 12. The chemical structure of the porphyrin ring.

Figure 13. Scheme of heme biosynthesis.

variety of toxic effects, such as vascular and intestinal spasms and neuropathologic changes (Marver and Schmid, 1972). It has been suggested that porphyrins may act to reduce the release of acetylcholine that usually occurs following ionic depolarization. Also, it appears that ALA (aminolevulinic acid), a porphyrin precursor, may inhibit brain tissue–dependent ATPase. Thus, it seems that intermediates in heme biosynthesis may be neurotoxins (Marver and Schmid, 1972).

Erythropoietic Porphyria

Erythropoietic porphyria is a metabolic disorder in erythroid cells of bone marrow which leads to excessive production of porphyrins. The disorder results in chronic skin photosensitivity and severe porphyrinuria associated with hemolytic anemia.

Hepatic Porphyria

The three hepatic porphyrias which are genetically transmitted (acute intermittent porphyria, variegate porphyria, and coproporphyria) have several common characteristics:

1. The acute phase has increased units of ALA and PBG (porphobilinogen); this is caused, in part, by increased activity of ALA synthetase (Marver and Schmid, 1972).

2. In many individuals, the metabolic disorder is in a latent form so that afflicted individuals are entirely free of symptoms or descriptions of nervousness. However, an acute attack may be precipitated by therapeutic doses of several drugs such as barbiturates, sulfonamides, general anesthetics, excessive amounts of ethanol, and chloroquine (Cripps and Curtis, 1962; Dean, 1963; Eales and Linder, 1968). These drugs are thought to induce hepatic ALA synthetase (Marver and Schmid, 1972).

3. Clinical symptoms are associated with similar neurologic features (Marver and Schmid, 1972).

4. Often the first clinical and biochemical signs of the disease occur during late puberty. This may be associated with naturally occurring steroids which induce ALA synthetase (Granick, 1966). Clinical exacerbations have also been noted in association with pregnancy (Vine et al., 1957; Neilson and Neilson, 1958; Gould et al., 1961; Petrie and Mooney, 1962) and the menstrual cycle (Perlroth et al., 1965; Marver and Schmid, 1972), which significantly alters steroid hormone levels.

Protoporphyria

Protoporphyria is a mild disorder characterized by chronic solar eczema caused by the overproduction of protoporphyrin IXa in liver and erythroid cells.

The gene frequency of the various porphyria metabolic disorders is uncertain; however, their occurrence is considered rare (Marver and Schmid, 1972). With regard to acute intermittent porphyria, the incidence is about 1.5 per 100,000 in Sweden (Marver and Schmid, 1972), Denmark (With, 1963), Ireland (Fennelly et al., 1960), and western Australia (Saint and Curnow, 1962) but considerably rarer in Negroes (Lyon, 1968).

A large outbreak of porphyria, affecting several thousand people, occurred during 1956 in Turkey. The toxic effects were caused by chronic ingestion of hexachlorobenzene, which was used as a fungicidal agent (Schmid, 1960; Cam and Nigogoysan, 1963). When the use of hexachlorobenzene was stopped in 1959, new cases of porphyria ceased. This disorder occurred in individuals who were not known to have any genetic predisposition toward developing porphyria (Cam and Nigogoysan, 1963).

Lead inhibits the activities of at least two enzymes (ALA dehydrase and ferrochelatase) concerned with the synthesis of heme in the red blood cells (Haeger-Aronsen, 1960; Schwartz et al., 1952; Lichtman and Feldman, 1963; Goldberg, 1968). Typical features of Pb toxicity are usually indicated by hypochromic anemia, increased free erythrocyte protoporphyrin, and excessive urinary excretion of ALA and coproporphyrin with near normal concentrations of PBG. Tschudy and Collins (1957) have also demonstrated that the herbicide 3-amino-1,2,4-trizole partially inhibits ALA dehydrase.

It is important to note that under normal conditions heme inhibits the activity of ALA synthetase, ALA dehydrase, and ferrochelatase. Thus, the scheme for heme biosynthesis operates under self-limiting controls (Marver and Schmid, 1972).

The presence of the genetic disorders which stimulate excessive activity of ALA synthetase or exposure to drugs, industrial chemicals, or hormones which induce this enzyme, along with Pb exposure, would be expected to produce higher than normal levels of ALA. In this instance, ALA is inhibited from being converted to heme by the Pb, thereby preventing the ALA from seeking its natural product (heme). This would result in excessive quantities of ALA being both excreted and accumulating in various tissues of the body. Since the production of heme would be reduced in such a situation, the activity of ALA synthetase would not be inhibited by negative feedback from the heme.

Cystic Fibrosis: A Predisposition to Respiratory Disease

Cystic fibrosis of the pancreas is a fairly common genetic disease of children and young adults in Caucasians of European origin. The clinical features result from a defect in exocrine gland function. Mucous secretions are often tenacious and thick, with extremely high levels of sodium and chloride characteristically present. The increased viscosity of mucus is an important cause of obstructive pulmonary disease in afflicted people (Lobeck, 1972).

The most important clinical features are both pulmonary and gastrointestinal (see "Ozone and Vitamin E"). Frequent pulmonary observations include progressive bronchiolus obstruction with complicating pulmonary infection, hyperaeration, and cor pulmonale. There is also a substance in plasma which causes ciliary dyskinesia (Spock et al., 1967).

Cystic fibrosis is seen clinically only in the homozygous state. Based on population surveys, it is estimated that 1 in 2000 Caucasians have CF. The incidence of heterozygotes in the Caucasian population is approximately 4 percent. Based on observations that parents of CF offspring do not exhibit an increased incidence of chronic pulmonary or gastrointestinal disease (Batten et al., 1963; Anderson et al., 1962; Hallet et al., 1965; Orzales et al., 1963), Lobeck (1972) concluded that heterozygotes for CF do not have clinical findings of CF. However, the literature contains various reports indicating that adult heterozygotes have exhibited "adult mucoviscidosis" expressed as pulmonary disease associated with somewhat elevated levels of sodium or chloride in sweat. It is important to note that chronic bronchitis may be associated with a tendency to higher sweat concentrations of sodium and chloride (Lobeck, 1972). Additionally, the presence of the ciliary dyskinesis factor in the serum of heterozygotes (Spock et al., 1967) further suggests the possibility of a predisposition to pulmonary disease in these individuals. Thus, it is expected that individuals homozygous and possibly those heterozygous for the CF trait are at increased risk to respiratory tract irritants such as ozone, sulfur dioxide, particulate sulfates, and numerous heavy metals.

Wilson's Disease and Hypersusceptibility to Vanadium

Mountain et al. (1953) reported that long-term feeding of vanadium (V) compounds to rats caused a reduction in the cystine content of rat hair. Subsequent research by Mountain et al. (1955) indicated that the fingernails of workers exposed to dusts containing V exhibited significantly lower amounts of cystine as compared with unexposed control populations. The decrease in cystine levels was correlated with the amount of V exposure. Mountain (1963), commenting on Keenan (unpub. obs.), indicated that, during studies of the V content of liver, there was found an inverse relationship between copper (Cu) and V levels.

Based on these observations of apparent antagonism between V and Cu, it was suggested by Mountain (1963) that certain toxic effects caused by V might be influenced by an induced Cu deficiency. The relationship between Cu and V was further clarified by the following observations and information provided in Table 10: (1) copper is an essential catalyst for keratinization (Marston, 1954), and (2) the increased neutral sulfur in the urine after rats ingested V (Mountain et al., 1959) is similar to the cystinuria related to Wilson's disease, where there is a genetically determined lack of normal protein for Cu (Stein et al., 1954). Table 10 indicates the similarity of the effects of V and Wilson's disease.

Based on this information, Mountain (1963) concluded that concurrent V exposure in a person with Wilson's disease may have at least additive and possibly synergistic effects with severe metabolic disturbances. This prediction is extremely difficult to confirm, because Wilson's disease (homozygous condition) is so rare as to effect less than 1 in 100,000 (Cooper, 1973). However, Mountain (1963), commenting on Leff and Klendshij (unpub. obs.), reports that such a case was found in which autopsy and other studies indicated that the genetic abnormalities were exacerbated by the environmental exposure to V. With regard to carriers (heterozygous condition) of the trait, it is estimated that there are 400,000 carriers in the United States. Particularly high incidences of the gene are found in Jews from Eastern Europe and non-Jews

Table 10.   Relationship of Vanadium to Wilson's Disease

Vanadium
1. Inverse relationship between liver vanadium and copper content.
2. Vanadium lowers cystine levels in rat hair following feeding.
3. Increased urinary neutral sulfur in rats fed vanadium.

Wilson's Disease
1. Lack of normal copper transport protein.
2. Low fingernail cystine.
3. Excessive excretion of urinary cystine (10 times normal).
4. Vanadium aggravates Wilson's disease.

Source.   Mountain (1963). Detecting hypersusceptibility to toxic substances. *Arch. Environ. Health.* **6:** 357. Copyright 1963, American Medical Association.

from the Mediterranean regions, especially Sicily (Greenblatt, 1974). Heterozygote carriers do not show clinical symptoms of the disease.

IMMUNOLOGICAL DISORDERS

Immunoglobulin A Deficiency and Respiratory Disease

Immunoglobulin A (Ig A) is the major immunoglobulin in saliva, tears, nasal, bronchial, and gastrointestinal secretions, and colostrum (Eisen, 1974; Koistinen, 1975). It is believed that the main function of Ig A is to act as a protective agent on secretory surfaces against foreign materials. In an effort to determine what, if any, clinical conditions are associated with Ig A deficiency, numerous studies have been performed on the possible role of Ig A deficiency in diseases in which the activity of the secretory surfaces may be disturbed. Also, among the various gastrointestinal disorders that have been reported in association with deficiency of serum IgA, malabsorption syndromes and pernicious anemia are the most frequent clinical diseases (Bjernulf et al., 1971; Douglas et al., 1970; Gelzayd et al., 1971; Ginsberg and Mullinax, 1970; Hermans, 1967; Mawhinney and Tomkin, 1971; Penny et al., 1971; Savilahti, 1973; and Wagner and Grossmann, 1969).

Several reports have indicated a direct relationship of patients with recurrent upper respiratory tract infections and the presence of Ig A deficiency (Ammann et al., 1970; Chia et al., 1970; Polmar et al., 1972; and Tushan et al., 1971). Infections in the lower respiratory tract together with Ig A deficiency have also been reported (Bachmann, 1965).

Reports of the frequency of selective Ig A deficiency in "normal" healthy populations varies from 1:300 to 1:3000 (Cassidy and Nordby, 1975; Collins-Williams et al., 1972; Frommel et al., 1973; Natvig et al., 1971; Pai et al., 1974; and Vyas et al., 1974). A recent doctoral dissertation reported an analysis of Ig A levels from all new blood donors (64,588) at the City of Helsinki Blood Donation Unit (Koistinen, 1975, 1975a). The results indicated a frequency of Ig A deficiency of 1:396. Other determinations made by the author, using different biochemical techniques, yielded results of a somewhat less frequent occurrence (e.g., 1:507, 1:661, 1:821). Thus, assuming a frequency of 1:400 in a population of 212 million (current United States population), more than 500,000 people have the Ig A deficiency (see Table 11).

Developmentally, cells with surface Ig A are detected during the fourth to fifth month of gestation. However, synthesis of Ig A by spleen cells does not become active until some weeks after birth. Considerable placental transfer of Ig molecules does occur with respect to Ig G, with only slight quantities of IgM and practically no Ig A molecules being transferred (Eisen, 1974).

Serum levels of the different Ig molecules reach "adult levels" at various times. For example, adult serum levels of Ig M are reached at about 10 months of age, Ig G at about 4 years, Ig A at about 9 to 10 years, and Ig E at about 10 to 15 years (Eisen, 1974). Figure 14 represents the relationship of immunoglobulin levels and the age of the typical human.

A report in 1972 of children 11 and 12 years of age in the Borough of the Bronx in New York City indicated that those who had lived in the heavily air-polluted Bronx for 5 to 12 years had 20 percent more lower respiratory infections than those who had lived in Riverhead (a town on eastern Long Island), which is relatively free of air pollution (Fosburgh, 1974). These findings may provide a partial understanding of the generally recognized greater susceptibility of children to respiratory infection.

Also, it has recently been reported that persons with a total lack of Ig A are more susceptible to influenza than others, including those with

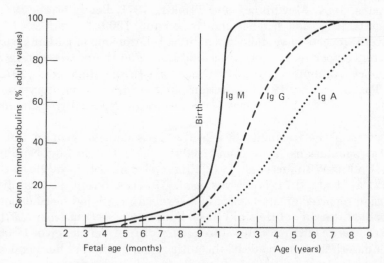

Figure 14. Maturation of serum Ig levels in man.

Source. Eisen (1974).

Table 11. The Frequency of Ig A Deficiency

Reference	Type of Group Tested	Number of Subjects	Frequency of IgA Deficiency
Koistinen (1975)	Blood donors	64,588	1:396
	Hospital patients	9,920	1:661
Bachmann (1965)	Healthy adults	6,995	1:700
Collins-Williams et al. (1972)	School children (5–19 years)	2,170	1:310
	Hospital patients (0–19 years)	7,261	1:458
Natvig et al. (1971)	Rheumatoid arthritis patients	3,187	1:398
	Blood donors	5,020	1:1,255
Frommel et al. (1973)	Blood donors	15,200	1:3,250

Source. Koistinen (1975*a*). Selective Ig A deficiency in blood donors. *Vox. Sang.* **29**:1. S. Karger AG, Basel, Switzerland.

partial deficiencies of Ig A (Koistinen, 1975*b*). However, it should be pointed out that an elevation of serum Ig G and Ig M concentrations has been observed in some of the Ig A–deficient blood donors. This may be an adaptational response to compensate for the Ig A deficiency. However, the role of the elevated serum Ig G is not clear. It is possible that these adaptations may contribute to the relatively good health of many of the Ig A–deficient blood donors (Koistinen, 1975*a*; Brandtzaeg, 1971; Brandtzaeg et al., 1968).

Genetic Control of Immunologic Hypersensitivity to Industrial Chemicals

Stokinger and Scheel (1973) have reported that the use of aliphatic and aromatic isocyanates has significantly increased. This is of importance, not only because of their direct toxicity to the worker, but also because of subsequent anaphylactic responses on reexposure of workers to the isocyanates (Figure 15). In other words, following inhalation of vapors of isocyanates which cause toxic symptoms, an asthmalike condition develops with a subsequent exposure to even trace levels of isocyanate. This is an "immediate" form of hypersensitivity similar to the allergy of

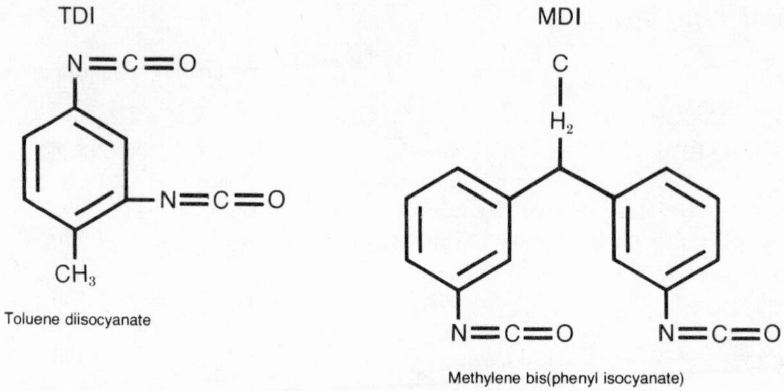

Figure 15. Several isocyanates used in plastic coating of foaming operations.

Source. Stokinger and Scheel (1973).

"hay fever." In other workers, a "delayed" response needs a much greater exposure to initiate it than the immediate reaction.

It has been further reported by Stokinger and Scheel (1973) that, from immunologic testing of more than 1000 sera, 0.5 percent of the worker population developed clinical cases of delayed hypersensitivity, and 1.5 percent developed clinical cases of immediate hypersensitivity with exposure. With regard to the remaining 98 percent, approximately 38 percent developed clinical symptoms when first exposed but became asymptomatic on continued exposure. The final 60 percent formed precipitating antibodies and developed immunity without symptoms.

Such studies clearly suggest that there is a highly variable immunological response in the worker population to isocyanates. It further implies that there is a genetic basis for the variation and that a certain segment of the population is predisposed (i.e., at increased risk) to the development of hypersensitive responses on isocyanate exposure. Knowledge of this human variation to isocyanate sensitivity has been used by industry to develop a highly successful preemployment screening program (see Stokinger and Scheel, 1973). Peters and Murphy (1970) have estimated that there are between 50,000 and 100,000 workers who are occupationally exposed to isocyanates in the United States. In addition to occupational exposure, consumers purchasing "instant polyurethane foam" kits for household use are also at risk of increased exposure. Peters and Murphy (1970) have indicated that household exposure to toluene diisocyanate vapor may exceed acceptable occupational levels.

OTHER GENETIC DISORDERS

Genetic Basis of Bronchogenic Carcinoma and Inducibility of Aryl Hydrocarbon Hydroxylase

It has been demonstrated that the metabolism of benzo(a)pyrene (BaP) can be measured in human lymphocyte cultures. Furthermore, there are genetic differences between individuals in their ability to metabolize BaP. It was additionally noted that individuals whose lymphocytes respond to inducers with high rates of BaP metabolism are more likely to be very susceptible to lung cancer on exposure to cigarette smoke (Kellermann et al., 1973).

It is thought that the majority of chemical carcinogens must be metabolically converted in specific tissues into the carcinogenic form. The polycyclic aromatic hydrocarbons are metabolized by certain mixed-function oxidases to certain reactive intermediates which can cause cell transformation, mutagenicity, and cytotoxicity. The activity of aryl hydrocarbon hydroxylase (AHH), one of the mixed-function oxidases, is under genetic control. The population frequency of this trait has been determined. The normal white population in the United States can be divided into three groups: low, intermediate, and high degrees of inducibility. Phenotype frequencies were 53 percent for low, 37 percent for intermediate, and 10 percent for high inducibility (Kellermann et al., 1973a).

By studying patients with bronchogenic carcinoma, it was found that groups with intermediate and high AHH inducibility had 16 and 36 times greater risk of lung cancer than the low group, assuming a risk of 1.0 in the low group (Kellermann et al., 1973a). The smoking habits of the patients with lung cancer revealed that all were heavy smokers with an average cigarette consumption of about two packages each day. Three did not smoke cigarettes; however, two smoked 10 to 15 cigars per day, and the other was a heavy pipe smoker. Duration of smoking ranged from 20 to 40 years.

These data support the hypothesis of Tokuhata (1964) that susceptibility to lung cancer is inherited. He indicated that familial clustering could be separated from environmental factors. His data showed that both heredity and cigarette smoking are important factors in lung carcinoma.

If the proposed relationship of AHH inducibility to bronchogenic carcinoma is more fully established by subsequent research, it may help to explain the generally accepted association of smoking and lung cancer incidence as well as the following anomaly: why some individuals may be heavy smokers for 40 to 60 years and never develop the least signs of lung cancer, but other individuals who may be only very light smokers may develop lung cancer. If we accept the AHH–lung cancer hypothesis, it becomes quite clear that certain segments of the population (i.e., those with a capacity for high induction of AHH) are at significantly increased risk to lung cancer as compared with others (i.e., those with a relatively low capacity to induce AHH). It would seem that the application of this knowledge may someday lead to screening clinics for prospective smokers. That is, if people intend to smoke but want to know their risk of developing lung cancer, they could have their lymphocytes tested to determine the degree to which they can induce AHH and thus predict to a certain extent their susceptibility to bronchogenic carcinoma via smoking.

Genetic Hypersusceptibility to Carbon Disulfide (CS$_2$)

Stokinger and Scheel (1973), commenting on Djuric et al. (1972), reported that there is human genetic variation in the ability to metabolize a CS$_2$-containing compound (tetraethylthiuram disulfide, TETD, disulfiram, antabuse) (Figure 16) one of the end products of which is CS$_2$. Individuals with a reduced capacity to metabolize TETD are considered genetically hypersusceptible. Evidence for the genetic basis for such susceptibility is based on the following information.

Following the ingestion of a 0.5-g tablet of TETD, urinary dithiocarb

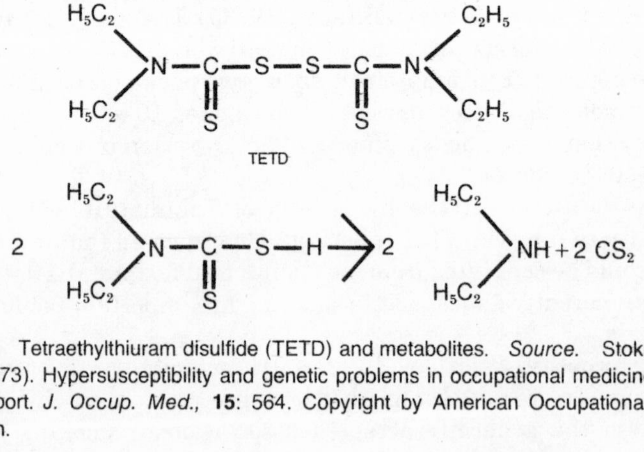

Figure 16. Tetraethylthiuram disulfide (TETD) and metabolites. *Source.* Stokinger and Scheel (1973). Hypersusceptibility and genetic problems in occupational medicine: A consensus report. *J. Occup. Med.*, **15**: 564. Copyright by American Occupational Medical Association.

excretion was determined in (1) 18 workers previously exposed to levels of CS_2 below the TLV or 60 mg/m³ (control group), (2) 21 workers exposed at higher levels but who had shown neither signs nor symptoms of CS_2 intoxication and were then labeled "resistant," and (3) 33 workers who exhibited polyneuritis or other symptoms of overexposure and had been removed from exposure and were then labeled "susceptible" (Stokinger and Scheel, 1973). Statistically significant differences between the resistant and control groups and between the resistant and susceptible groups were noted with regard to metabolized TETD. It can be reasonably concluded that the evidence for the genetic basis for this susceptibility is based on only very preliminary evidence. Considerably more information is needed before firm conclusions can be drawn.

Genetic Control of Chloroform Toxicity

Chloroform, a widely used substance in medicinal and industrial applications, has been found to damage the liver and kidney in man (Von Oettingen, 1964) and experimental animals (Drill, 1952). Long-term chloroform exposure causes liver tumors in mice (Eschenbrenner, 1944), and nonanesthetizing exposures to adult female rats may be toxic to the fetus (Schwetz et al., 1974; Thompson et al., 1974).

Accidental exposure of various mouse strains to chloroform vapor has indicated that response varies from one strain to another (Bennet and Whigham, 1964; Hewitt, 1956; Jacobsen et al., 1964). Recent experiments by Hill et al. (1975) demonstrated that genetic determinants control chloroform toxicity in mice and that the level of toxicological activity is directly related to the amount of renal accumulation of chloroform. More specifically, their results showed an intermediate or multifactorial genetic control of chloroform-induced renal toxicity and death.

Eschenbrenner (1944) has reported that males and females of the same strain show similar thresholds to hepatic damage. Deringer et al. (1953) reported sex differences with respect to renal toxicity; that is, females die of chloroform-caused hepatic lesions before developing renal damage, but males are quite susceptible to renal damage.

Hill et al. (1975) note that sex hormones have been implicated in the sex difference in renal toxicity. For example, immature male mice and castrated adult mice are resistant to chloroform-induced damage but are sensitized by testosterone; males administered estrogen are resistant. Females administered testosterone become sensitive to renal toxicity. The authors suggested that mouse strain variation in androgen synthesis and activity may explain the strain differences in their LD_{50} studies.

Hill et al. (1975) indicated that the capacity to metabolize many commonly used drugs by humans is genetically determined (see Vesell and Page, 1968, 1968a, and Alexanderson et al., 1969). They suggested that the ability to metabolize chloroform may also be under genetic control in humans as in their animal models. If this is found to be the case, one could expect the existence of human high-risk groups on the basis of different capabilities to detoxify and excrete chloroform. Since chloroform is now known to be present in the drinking water of more than 150 American cities (Science News, 1976), there is an important need to investigate whether specific segments of the human population may be at increased risk to the toxic or carcinogenic effects of chloroform.

Sulfite Oxidase Deficiency: A Factor in SO_2, Sulfite, and Bisulfite Toxicity?

Hickey et al. (1976) have recently developed an interesting, yet highly controversial (see Alarie, 1976) theoretical prediction that toxicity from SO_2 and dietary sulfites (food preservatives) may be exacerbated by the occurrence of a sulfite oxidase deficiency. This association is based on studies which indicated that SO_2 and sulfite may be mutagenic and that the enzyme sulfite oxidase is capable of detoxifying the mutagen sulfite to the sulfate form.

A brief survey of possible cancer and SO_2 associations reveals that (1) LX mice that were exposed to SO_2 had a greater although statistically insignificant prevalence of lung tumors than control specimens (Peacock and Spence, 1967), (2) studies in the United States have also associated SO_2 levels with the incidence of several types of cancer (Hickey et al., 1970), and (3) mutagenic activity by SO_2 in Tradescantia (twice the spontaneous rate) has been reported (Sparrow and Schairer, 1974). With regard to the toxic effects of sulfite, research has indicated that bisulfite is able to induce a variety of nucleotide base changes in the DNA of microorganisms (Shapiro et al., 1974). However, the mutagenic activity at physiological pH is only 10 percent of the mutagenic activity at the optimal pH. Further, based on bacteriophage T4 studies, Summers and Drake (1971) concluded that bisulfite was a moderately strong mutagen.

As previously indicated, sulfite oxidase is known to catalyze the oxidation of sulfite to sulfate. Also, elevated levels of sulfite and bisulfite have been noted in a human patient with the sulfite oxidase defiiency (Mudd et al., 1967). Hickey et al. (1976) indicated that individuals with the enzyme deficiency would be at increased risk to the mutagenic activity of bisulfite and SO_2. The mutagenic activity of SO_2 would seem to be an indirect one, that is, by influencing the formation of sulfite. Yet Hickey et al. (1976) did not present any original or referenced data to support such an occurrence. Also, it is important to point out that the deficiency has been established in only one infant (Mudd et al., 1967). However, Hickey et al. (1976) contend that structural polymorphisms of sulfite oxidase may be responsible for subtle changes in enzyme activity and thus the development of differential susceptibility. Consequently, at this point, more information is needed concerning the genetic frequency of the defect and the occurrence of enzyme polymorphisms, including their functional differences and frequency in the population. However, even more important is the research to test the hypothesis of Hickey et al. (1976) with appropriate animal models.

Leber's Hereditary Optic Atrophy and Cyanide Toxicity

In 1871, Leber reported the occurrence of a rare genetic disorder in which those affected experience visual failure either subacutely or insidiously. The condition often finally results in more or less severe bilateral central scatomata and pallor of the optic discs.

The condition is usually first recognized in males in late teenage years or early twenties, although the range of onset is quite wide. In Western cultures only about 15 percent of those with Leber's optic atrophy are women (Wilson, 1965). Based on the hereditary nature of the disease and clinical phenomena, Wilson (1965) suggested that the disease is caused by an inborn metabolic error whose clinical expression is significantly affected by exogenous (environmental) factors; smoking is considered one such factor. The association between smoking and Leber's optic atrophy was suggested by Wilson (1965a) because of the recognized presence of several neurotoxic agents in cigarette smoke, including cyanide, as well as the knowledge that repeated exposure of cyanide to experimental animals has produced neurological effects similar to those of Leber's optic atrophy.

Normally, cyanide is highly reactive and is rapidly metabolized to less toxic compounds. Consequently, a significant proportion of cyanide is quickly converted to thiocyanate and excreted. With this considered, Wilson (1965a) experimentally tested the hypothesis that smoking may adversely affect those with Leber's optic atrophy. The results demonstrated that patients with the disease had significantly lower levels of thiocyanate in the plasma and urine than did the controls when both groups were smoking. The author suggested that such patients may have an inborn metabolic error with respect to cyanide metabolism and may be unable to detoxify cyanide to thiocyanate sufficiently.

Such evidence certainly implies that individuals with Leber's optic atrophy may be at increased risk to cyanide poisoning. However, the results are certainly of a preliminary nature, and considerably more research is needed before any reliable generalizations can be made. Presently, research on the gene frequency of this trait is necessary.

The toxicity of cyanide is also thought to be markedly influenced by the nutritional status of the individual. As previously mentioned, cyanide is thought to be detoxified, in part, by its metabolic conversion to thiocyanate. This detoxification pathway involves substrates derived from cysteine. It is significant to note that epidemiological studies involving 320 Nigerian patients (14 to 24:1000 adults) with "tropical neuropathy" quite similar to Leber's disease, have revealed that the geographical distribution of the tropical neuropathy is closely associated with the consumption of cassava, a tuber whose outer integument has high levels of the cyanogenic glycoside linamarin. Furthermore, the diet of the 320 Nigerians was found to be generally quite low in cysteine and other sulphur-containing amino acids. It was suggested that the combination of low dietary levels of these amino acids and their increased use for the detoxification of cyanide may be the cause of the severely depleted levels of plasma cysteine and methionine in these patients. However, the mode of action by which toxic symptoms are affected is not precisely known, although it may include direct intraneuronal enzymic inhibition by cyanide or by the inactivation of vitamin B_{12} (Anonymous, 1969).

Genetic Susceptibility to Ultraviolet-Induced Skin Cancer

Considerable controversy has focused on the effects of ultraviolet radiation on human health as a result of studies which have indicated that supersonic jets (e.g., Concorde) and spray propellants (e.g., chlorofluoromethanes) may be affecting a reduction in the stratospheric ozone layer. The ozone layer absorbs potentially harmful wavelengths of ultraviolet radiation (250 to 320 nm), thereby preventing such radiation from reaching and adversely affecting human life on earth. It is generally accepted that the formation of the ozone layer during our geological past was a necessity for the emergence and evolution of terrestrial life. However, despite the efficiency of the shielding activity of the ozone layer, a biologically important amount of ultraviolet radiation does ultimately reach the earth's surface, causing both beneficial and harmful effects on humans.

Beneficial Aspects of Ultraviolet Radiation

Ultraviolet radiations (280 to 320 nm) are known to penetrate the skin and have been reported to effect the conversion of provitamin D to vitamin D. Since vitamin D is present in large amounts in only a few foods (e.g., bony fishes), the formation of vitamin D via the action of ultraviolet radiation has been suggested as playing an important role in acquiring sufficient amounts of vitamin D for proper development. Without adequate quantities of vitamin D in the diet, proper development of skeletal tissue would be prevented, and children would become rachitic. Conversely, the presence of excessive vitamin D can result in a toxic condition known as hypervitaminosis D, which is characterized by calcification of soft tissues (e.g., kidneys). To regulate effectively the amount of ultraviolet radiation which penetrates the skin, humans have evolved a melanin pigment system which can effectively screen out ultraviolet radiation or permit penetration, depending on the degree of pigmentation. Black skin is considerably more efficient in preventing penetration of ultraviolet radiation than lighter skin. According to evolutionary theory, skin color, including the phenomenon of reversible tanning, is a physiological adaptation to regulate the level of vitamin D in the body by controlling the ultraviolet-mediated synthesis of vitamin D (see Loomis, 1967, for a detailed discussion of the skin color–vitamin D hypothesis). Calabrese (1977a) has suggested that elevated levels of "breathable" ozone in our urban areas may actually act to increase our daily dietary requirement of vitamin D by absorbing ultraviolet radiation which would usually reach the body and lead to the formation of vitamin D.

Carcinogenic Activity of Ultraviolet Radiation

In addition to being involved with the synthesis of vitamin D, considerable evidence indicates that ultraviolet radiation has carcinogenic activity. The evidence associating ultraviolet radiation with skin cancer has a relatively long history, stretching back to several reports in the 1890s (Unna, 1894; Dubreuilh, 1896; and Shield, 1899) and continuing up to the present (Urbach, 1975). During this time, the evidence of the ultraviolet–skin cancer association has been derived from complementary areas of research, including animal studies and human epidemiological and clinical observations. A short summary of the evidence implicating midultraviolet (250 to 320 nm) as a significant factor in the development of most human skin cancer is as follows (see Urbach, 1975):

1. Skin cancer appears on parts of the body most exposed to sunlight, especially the head, neck, arms, and hands (Anchev et al., 1966; Silverstone and Gordon, 1966; and McGovern, 1966).
2. Squamous cell and basal cell carcinomas occur primarily on skin locations most often exposed to solar radiation (Urbach, 1966; Allison and Wong, 1967).
3. Anchev et al. (1966) and Blum (1959) reported a significantly greater incidence of skin cancer in outdoor workers (e.g., farmers).
4. For similar skin types the rate of skin cancer increased dramatically as the latitude decreased, especially within the middle latitudes

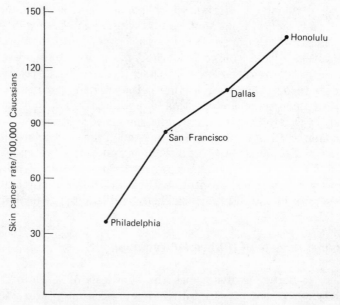

Figure 17. Skin cancer annual incidence rates per 100,000 white population, Honolulu County, 1955 and 1956; selected mainland areas, 1947 and 1948. *Source.* Allison, S. D., and Wong, K. L. (1967). Skin cancer: Some ethnic differences. In: *Environments of Man.* Addison Wesley Publishing Co., Reading, Mass., p. 69. Copyright 1957, American Medical Association.

(Urbach, 1975). Figure 17 which strongly supports the work of Urbach (1975) represents the annual skin cancer incidence rates in Honolulu County and selected mainland areas. Scotto et al. (1974) have indicated that the Third National Cancer Survey revealed that by age 50, 1 percent of all male Caucasians in the area around Dallas, Texas, developed skin cancer malignancy. By age 80, the figure was 3 percent. These rates were twice those of individuals from Minneapolis–St. Paul.

5. Specific racial groups, especially those of Celtic origin, are more susceptible to the development of skin cancer than other whites. The most susceptible individuals have skin with little pigment, scattered freckles, light-colored hair, blue or gray eyes, and a facility to sunburn (O'Beirn et al., 1970; Urbach et al., 1972). Negroes and Orientals are generally much less susceptible than Caucasians (Oettle, 1966; and Allison and Wong, 1967). In fact, skin cancer is 45 times more common among Caucasians as compared with non-Caucasians in Honolulu, Hawaii. Figure 18 shows the skin cancer incidence rates by race in Honolulu County.

6. Finally, animal studies, most often using mouse skin, have repeatedly demonstrated that skin cancer can be produced by repeated exposure to wavelengths between 250 and 320 nm, with 280 to 320 nm the most carcinogenic (Epstein, 1966; Blum, 1948).

Health Implications of a Reduction in the Ozone Layer

Based on the previously summarized data concerning the carcinogenic effects of ultraviolet radiation, much public and scientific commotion has resulted from the possible depletion of the ozone layer.

It has been estimated that supersonic transports which cruise in the stratosphere would reduce the ozone layer by 0.07 percent for a small fleet (NAS, 1975a) and by 50 percent for a fleet of 500 (Johnson, 1973). Depletion of the ozone layer by chlorofluorocarbons has been projected to be 16 percent by the year 2000 if usage of these propellant chemicals were to increase at 10 percent per year (Wofsy et al., 1975). A recent National Research Council (NRC) report has concluded that the continued release of halocarbons at the 1973 rate will ultimately produce between a 2 and 20 percent reduction in stratospheric ozone, with the probable reduction about 7 percent (Maugh 1976). Schultze (1974) has calculated that if the ozone concentration in the stratosphere were diminished by 10 percent, the total ultraviolet radiation dose would be increased by 19 percent in the middle latitudes (40°N to 60°N) and by 22 percent in the equatorial zones. Fears et al. (1976) have developed a mathematical model to predict the relative increases in skin cancer incidence and mortality associated with Schultze's projected changes in ultraviolet levels (erythema dose). Specifically, they have predicted an increase by 15 to 25 percent in the melanoma incidence of males, depending on the degrees north latitude. Similar effects are predicted for females.

Individuals at High Risk to Ultraviolet Radiation

One of the important recommendations of the NRC's report was to undertake research to identify population groups with a drastically higher than normal susceptibility to malignant melanoma. Previously cited research has noted the marked enhancement of ultraviolet radiation–induced skin cancer in Caucasians, especially those with fair complexions. However, there are other groups, although less numerous,

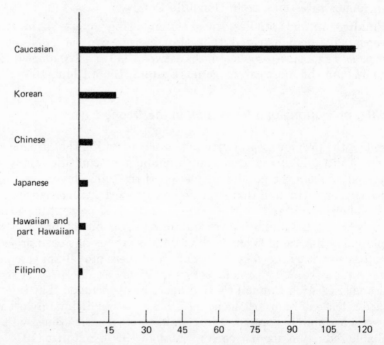

Figure 18. Skin cancer incidence rates per 100,000 population by race, Honolulu County, 1955 and 1956. *Source.* Allison, S. D., and Wong, K. L. (1967). Skin cancer: Some ethnic differences. In: *Environments of Man.* Addison Wesley Publishing Co., Reading, Mass., p. 69. Copyright 1957, American Medical Association.

who should also be considered at increased risk to skin cancer from the ultraviolet radiation. These groups include people with birth marks such as moles (Oettle, 1966; McGovern, 1966), the elderly, especially those who develop Hutchinson's melanotic freckle, a lesion present in skin severely affected by the sun (Anchev et al., 1966; Oettle, 1966), oculocutaneous albinism, including both tyrosinase positive and negative phenotypes and other reduced melanin conditions (discussed below), and individuals with xeroderma pigmentosum.

Albinism is a genetically transmitted disorder of the melanin pigment (melanocyte) system which is recognized by dramatic reduction of melanin in the skin, hair, and eyes. Albinism occurs not only in man but also in other mammals, as well as in birds, reptiles, amphibians, and fish. Melanocytes are specialized cells where melanin synthesis occurs. In man, mature melanocytes are usually found in definite regions: skin (hair bulbs, dermis, and dermoepidermal junction), mucous membranes, nervous system (piarachnoid), and eye (uveal tract and retinal pigment epithelium). The melanin pigment is synthesized in the cytoplasm of melanosomes by the enzymatic conversion of its precursor (tyrosine) by tyrosinase. With the exception of those located in the hair bulbs and retinal pigment epithelium, all melanocytes seem to have the ability to develop into malignant melanomas. With respect to albinism, the metabolic disorder may affect the entire melanocyte system (oculocutaneous albinism) or the melanocytes at a specific site (ocular albinism) (Fitzpatrick and Quevedo, 1972). Oculocutaneous albinism is known to occur in the different races of man, with the affected having significant reduction of melanin in the hair, skin, and eyes. There are two phenotypic expressions of oculocutaneous albinism that are distinguished by either the presence or absence of tyrosinase and are thus referred to as tyrosinase positive or tyrosinase negative (Fitzpatrick and Quevedo, 1972).

Associated with the reduction of the melanin pigment is an enhanced susceptibility to the toxic effects of solar radiation. Individuals with albinatic skin are known to have a high frequency of solar keratosis and basal cell and squamous cell carcinomas of the skin of the exposed area (Curban, 1951; Shapiro et al., 1953). Malignant melanomas are also known to occur in these individuals (Young, 1958; Oettle, 1966; Garrington et al., 1967).

The incidence of tyrosinase-negative oculocutaneous albinism in the Negro and Caucasian population is approximately 1:34,000 to 36,000 persons (Witkop, 1972). Tyrosinase-positive oculocutaneous albinism has an incidence of approximately 1:14,000 in Negroes and 1:60,000 in Caucasians (Witkop, 1972). Oculocutaneous albinism is present with a high incidence among various American Indians: Tule Cuna Indians, 1:143; Hopi, 1:227; Jemez, 1:140; Zuni, 1:247; Navajo, 1:3,750 (Witkop, 1972). Furthermore, several investigators have reported that the total incidence of oculocutaneous albinism in Ireland is approximately 1:10,000 to 15,000, with an estimated mutation frequency for albinism of 3.3×10^{-5} to 7.0×10^{-5} per gene per generation (Froggatt, 1957; Witkop, 1972). Interestingly, it has been noted that Ireland has the third highest skin cancer rate in the world, despite the fact that the sun shines only 30 percent of the possible sunshine hours (MacDonald, 1966).

According to Herndon and Freeman (1976), patients with vitiligo (a skin disease manifested by smooth, milk-white spots on various parts of the body), phenylketonuria (PKU) or localized loss of pigmentation because of injury or inflammatory skin disease also represent groups at high risk to skin cancer. With regard to PKU, the frequency of carriers in the general population is 1:80, and the frequency of the afflicted PKU individual is 1:25,000 (Knox, 1972). The major biochemical changes in PKU are related to the accumulation in the tissues of that part of the dietary L-phenylalanine which would normally be converted to tyrosine. No harmful effects have been reported in the heterozygote as yet, despite the fact that the serum of heterozygotes may have elevated phenylalanine levels. To investigate whether carriers of PKU are more susceptible to ultraviolet-induced skin cancer is especially relevant when one considers that there are greater than 2.5 million PKU carriers in the United States.

In order to survive the daily exposure to appreciable amounts of ultraviolet radiation, plants and animals, including man, have evolved DNA repair processes to protect affected cells and to assist in the recovery from the damaging radiation effects. Urbach (1975) has reported that three major kinds of repair processes have been described:

1. The damaged molecule may be repaired to its active condition *in situ*. This is achieved either by an enzyme-catalyzed repair or by "decay" of the damage to some inconsequential (or harmless) form.
2. The damaged part can be excised and replaced with undamaged components which restore normal function.
3. The damage may not be specifically repaired, but the cell may either bypass the defect or make some other functional adaptation.

A genetic defect in DNA repair processes has been reported by Cleaver (1968) in cultured cells from ultraviolet radiation–sensitive skin cancer–prone patients with the condition xeroderma pigmentosum. Usually, xeroderma pigmentosum fibroblasts are extremely sensitive to ultraviolet light and perform reduced amounts of repair replication during repair of damage to DNA. Subsequent findings by Cleaver and Carter (1973) reported the existence of genetic variants of xeroderma pigmentosum with regard to their different abilities to repair ultraviolet damage to the cells. This phenomenon is thought to be of potential significance for the hypothesis that genetic background may be an important variable in one's susceptibility to developing skin cancer. It is of interest to note that a variety of cell types, including liver, kidney, and brain, have excision repair mechanisms which may assist against the clastogenic effects of systemic mutagens.

According to Urbach (1975), in mouse skin and in most cancer patients, the DNA repair process does not appear to be impaired. Consequently, he concluded that the lack of DNA repair cannot be considered the basis of most skin cancers. Commenting on the research of Epstein et al. (1971) and Zajdela and Latarjet (1973), he suggested that the development of skin cancer by ultraviolet radiation is actually initiated by repair of DNA, permitting the cell to survive, yet facilitating the occurrence of errors in DNA replication resulting in an enhanced likelihood of malignant change. This, of course, implies a defect in the repair system also. However, the defect would be more subtle and difficult to detect. It would seem that the discovery by Cleaver and Carter (1973) of the existence of genetic variants with respect to xeroderma pigmentosum and their variable DNA repair capabilities clearly underlines the potential differential susceptibility of the general population if genetic polymorphisms for repair processes do exist and are widespread. At this point, further research is needed to clarify the role of DNA repair mechanisms in carcinogenesis before any sound generalizations can be made.

Noncarcinogenic Skin Irritation by Ultraviolet Radiation

Lupus Erythematosus

It is well known that sunlight adversely affects the skin of those afflicted with both systemic and cutaneous lupus. Dubois (1974) has reported that approximately 25 to 50 percent of patients with systemic lupus erythematosus (SLE) have adverse responses to typically tolerated levels of ultraviolet radiation and nearly twice this percent of patients with discoid LE have adverse effects. Herndon and Freeman (1976) have reported that irritation of existing skin lesions, fever, and progression of systemic activity may occur following excessive exposure; however, sunlight is not considered an integral component of the pathogenesis of LE. They further indicate that numerous SLE patients tan easily and may even sustain sunburn without relapse. However, individuals with LE are usually well advised to avoid overexposure.

Photoallergic Drug Reactions

Herndon and Freeman (1976) have reported that ultraviolet radiation interacts with various chemical substances to form complete photo-antigens which may cause anaphylactic responses in sensitive individuals. Examples of such photoallergic substances include halogenated antiseptic compounds used in soaps and cosmetics. Freeman et al. (1970) have reported that during typical usage of these photoallergic substances sensitive individuals, even those of dark-skin racial background, became susceptible to painful erythema four times more quickly (5 min compared with 20 min) than controls in Texas noonday sun. The frequency of these hypersensitive individuals in the general population remains to be determined.

Nutritional Deficiencies and Supplementation

Considerable interest has focused on the role of dietary factors as modifying influences to the toxic effects of various environmental and occupational pollutants. Vitamins A, B, C, D, and E as well as numerous minerals, including Ca, Fe, Mg, Mn, P, Se, and Zn, have been reported to reduce the toxic effects of particular pollutants. This should not be particularly surprising, since it is well known that a healthy organism is better equipped to ward off the onset of various stressing agents such as bacteria, viruses, and, as indicated above, the effects of environmental pollutants.

The health status of any individual can be understood by examining the interaction of one's adaptive capacity versus the environmental stressors. It is obvious that those individuals who have outstanding adaptive capacity as well as minimal environmental stress are best capable of experiencing good health, and the reverse is also true. Consequently, a wise approach toward ensuring good health is both the strengthening of one's adaptive capacity and the reduction of environmental threats or stressors.

Presently, strong attempts have been

consciously made via environmental and occupational health standards to reduce significantly the level of stressing agents such as ozone, sulfur dioxide, carbon monoxide, and other pollutants. However, these standards have not given sufficient attention to the other half of life's equation, that is, man's adaptive capacity.

Even though the alteration of one's genetic make up and ontogenic development is not presently accessible to human control, the improvement of our daily dietary habits as a result of improved nutrition education programs as well as nutritionally balanced lunches for school children and the elderly can be used to make substantial improvements in man's adaptive capacity. Various researchers, in fact, are now conducting preliminary studies on the supplementation of normal human diets by various substances, including certain vitamins and minerals, to determine if the toxic effects of certain pollutants can be alleviated. Consequently, since our ability to improve our adaptive capacity is realistically limited to this one area, the opportunity to make practical contributions to the quality of human life via nutritional improvements should be aggressively pursued.

VITAMIN A DEFICIENCY AND POLLUTANT EFFECTS

Biochemical research during the last two decades has indicated several important associations between dietary vitamin A levels and the toxic effects of various pollutants. The following section discusses how vitamin A deficiency and supplementation may affect the toxicity of some chlorinated hydrocarbons and environmental carcinogens.

Interaction with PCBs and Chlorinated Hydrocarbon Pesticides

Vitamin A storage in rat liver was reduced by as much as 50 percent when these animals were exposed to Aroclor 1242, a polychlorinated biphenyl. If only marginal amounts of vitamin A were in the diets of such rats, it was thought that avitaminosis could result. This suggested an alteration of lipid metabolism and preferential absorption from the gastrointestinal tract resulting from PCB exposure. Such a study further implies that humans who are deficient, to various degrees, or marginal in vitamin A levels may have their deficiency exacerbated with concomitant exposure to PCBs (Cecil et al., 1973). Other studies have indicated that both DDT and dieldrin, like Aroclor 1242, also reduce liver stores of vitamin A. Addition of methionine, an amino acid, to the diet reduces this effect of DDT. It is thought that methionine assists in the absorption of vitamin A from the intestinal mucosa. The increased toxicity of endrin to rats on a plant protein diet may be explained by the methionine deficiency of soy protein (Tinsley, 1969; Shakman, 1974).

Tumor Development

Saffiotti et al. (1967) have demonstrated that the administration of vitamin A, following exposure of the respiratory tract to benzo(a)pyrene (BaP), can inhibit the development of squamous tumors in the tracheobronchial mucosa of rats.

Chu and Malmgren (1965) reported that the occurrence of squamous cell carcinomas from the uterine cervix and vagina of hamsters topically painted with 7,12-dimethylbenz(a) anthracene (DMBA) was inhibited when 10 percent vitamin A palmitate was added to the olive oil used as vehicle for the carcinogen. Chu and Malmgren further reported that the induction of squamous cell tumors in the esophagus and the forestomach of hamsters fed DMBA or BaP was also inhibited when vitamin A palmitate was added to the carcinogen. Other research by Davies (1967) indicated a higher regression rate of skin papillomas in rhino mice painted with DMBA when these mice also were given a dietary supplement of vitamin A. Finally, Rowe and Gorlin (1959) reported that in the hamster cheek pouch, which is lined with a thin squamous epithelium, vitamin A deficiency results in a higher incidence of squamous tumors following topical application of DMBA.

Additional studies summarized by *Science News* (1976) have provided further support to the vitamin A–cancer prevention relationship. For example, a study of 8000 men indicated that a relatively low dietary intake of vitamin A correlated with a relatively high incidence of lung cancer after matching the men for similar smoking habits. Further research by Michael B. Sporn, of the National Cancer Institute, has demonstrated that retinoids can prevent precancerous tissues from becoming cancerous in both *in vitro* and *in vivo* animal studies. Also, Paul Newberne et al., of MIT, have reported that providing rats with 10 times their normal vitamin A intake dramatically reduced their susceptibility to lung cancer.

Although surveys have shown that it is not uncommon for the diet to fall slightly below the recommended allowance in the United States for vitamin A, marked symptoms of vitamin A deficiency are not often seen (Bogert et al., 1973). However, in the diets of teen-age girls, pregnant women, and older persons, the level of vitamin A intake may often be too low to allow for a proper margin of safety (Bogert et al., 1973). Furthermore, the average amount of vitamin A available to each person per day in the United States has decreased from a level of 8700 IU in 1947–1949 to the present level of 7800 IU (Friend, 1970). The National Nutritional Survey in New York City reported that 46 percent of children under 7 years old in lower-income families were found to have vitamin A deficiency. Eighteen percent of children of the same age in upper-income families were deficient. Finally, 27 and 25 percent of children (7 to 12 years of age) from lower- and upper-income families, respectively, had vitamin A deficiencies. The dividing line between lower- and upper-income families was placed at an annual income of $4158 for a family of four (Anonymous, 1971).

It should be noted that amounts of vitamin A in the body can reach excessive levels and cause intoxication in man. Chronic hyper-vitaminosis A has occurred in patients who received large doses of vitamin A for dermatological reasons without supervision and faddists who ingest extremely large doses in the diets (Smit and Hofstede, 1966; Ammann et al., 1968; Oliver, 1958; DiBenedetto, 1967). See Roels (1966) for a review of the role of vitamin A in health.

Deficiencies of riboflavin, one of the B vitamins, have also been found to enhance the carcinogenicity of polycyclic aromatic hydrocarbons in similar fashion to the vitamin A deficiency condition described above (Wynder and Klin, 1965; Wynder and Chan, 1970, 1972; Wynder et al., 1975). They indicated that riboflavin-deficient-fed mice are more susceptible to the carcinogenic effects of DMBA as compared with mice fed with normal riboflavin levels. Figure 19 illustrates this relationship.

VITAMIN C DEFICIENCY AND POLLUTANT TOXICITY

Vitamin C (ascorbic acid) deficiency has been found to potentiate the effects of several pollutants, while vitamin C supplementation may prevent the harmful effects of some pollutants. The following information summarizes how various levels of vitamin C may modify the toxic effects of 10 commonly encountered pollutants (Stone, 1972).

Inorganic Metals, Metaloids, and Nonmetals

Arsenic

The treatment of syphilis has involved the use of many chemicals, including various arsenical compounds. In conjunction with its anti-

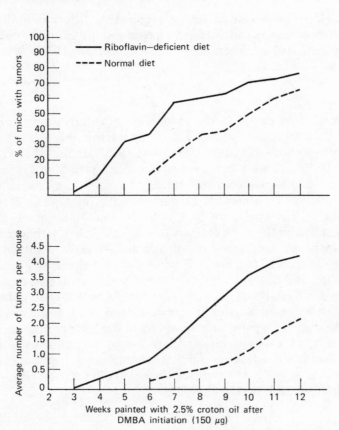

Figure 19.   Skin tumor yield in riboflavin-deficient mice (Swiss ICRQ).   *Source.*   Wynder et al. (1975).

bacterial effects, the arsenicals have also caused toxic effects in various patients. It has been shown that treatment of patients with ascorbic acid prevents the toxic effects of As (Abt, 1942; Lahiri, 1943; McChesney et al., 1942; McChesney, 1945; and Marocco and Rigotti, 1962).

Cadmium

Ascorbic acid is also known to interact with Zn and Cd. In quail, dietary ascorbic acid supplementation prevents Cd-induced anemia and growth retardation (Fox and Fry, 1970).

Chromium

Vitamin C, as a dermal treatment application, reduces hexavalent chromate to the nontoxic trivalent form. It has been suggested that if vitamin C is incorporated into a respiratory filter, it may reduce respiratory cancer in industrially exposed groups (Samitz et al., 1962, 1968; Samitz, 1970; Pirozzi et al., 1968; Shakman, 1974).

Lead

Vitamin C is essential to the production of intercellular connective tissue in the muscles of animals. Lead apparently inhibits the reserve capacity to produce this tissue (Aaronson, 1971). Holmes et al. (1939), in a study of 400 workers in an industrial plant with high Pb exposure, reported that the symptoms of chronic Pb poisoning resembled subclinical scurvy. Treatment of 17 workers with 100 mg/day of ascorbic acid relieved the symptoms in 1 week or less. Marchmont-Robinson (1941) confirmed the work of Holmes et al. (1939) in another industrial study with 303 employees of an automobile body plant where Pb exposure in the air and dust was quite high. Furthermore, Pillemer et al. (1940) found that high levels of ascorbic acid protected guinea pigs from the toxic effects of Pb. More specifically, only 2 of 26 guinea pigs exposed to 50 mg/kg of ascorbic acid developed paralysis or spasticities, and none died of Pb poisoning; however, in the low-ascorbic-acid group (2.5 mg/kg), 18 of 44 animals developed some form of neuroplumbism, and 12 died of Pb poisoning. Additional animal studies indicated that ascorbic acid significantly reduced Pb toxicity in tadpoles (Han-Wen et al., 1959) and rabbits (Uzbekov, 1960).

Mercury

Vauthey (1951) reported that a particular dose of mercury cyanide injected into guinea pigs killed 100 percent of the specimens within an hour of the exposure. However, if the guinea pigs were given megascorbic levels of ascorbic acid (equivalent to 35 g/day for a human weighing 70 kg) before the Hg treatment, 40 percent survived the Hg treatment. A similar protective effect was reported by Mavin (1941) and Mokranjac and Petrovic (1964) when mercury dichloride ($HgCl_2$) was used as treatment.

Various mercurial compounds are used as diuretics to reduce body fluids. Along with beneficial medicinal effects, some patients experience certain types of toxicity. Chapman and Shaffer (1947) have demonstrated that the toxic effects of certain mercurial diuretics could be reduced by prior or simultaneous administration of ascorbic acid.

Nitrates and Nitrites

Individuals suffering from methemoglobinemia are occasionally treated with vitamin C, since it reduces the MetHb to hemoglobin. Both animal and human research support the beneficial action of vitamin C in individuals with excessive MetHb. For example, when vitamin C–deficient pregnant guinea pigs were injected with nitrite, a high incidence of abortion or fetal death occurred due to MetHb (Kociba and Sleight, 1970). In Hungary, large doses of vitamin C administered to infants have significantly lowered the incidence and severity of MetHb toxicity (Török, 1971).

Organic Compounds

Benzene

Human exposure to benzene has been shown to deplete the body stores of vitamin C and to bring on a condition of subclinical scurvy. The administration of vitamin C aids in the prevention of the symptoms of chronic benzene exposure (Thiele, 1964). Daily ingestion of orange juice prevented chronic benzene poisoning in a South African chemical plant (Lurie, 1965). In agreement with the previous research, Forssman and Frykholm (1947) indicated that exposure to benzene causes an increased need of ascorbic acid and that a supplemental amount of ascorbic acid provides increased resistance to the effects of benzene.

Insecticides: Organochlorines

Organochlorine insecticides such as DDT, dieldrin, and lindane are usually *in vivo* substrates of microsomal enzymes such as the mono-oxygenase and conjugating systems. Insecticide metabolism and toxicity are affected by a variety of nutritional factors, including vitamin C (Street and Chadwick, 1975; Zannoni and Lynch, 1973; Kato et al., 1969). Numerous organochlorine insecticides are inducers of the liver microsomal enzymes. Their inductive activities result in the stimulation of the glucuronic pathway, causing increased ascorbic acid biosynthesis in the rat (Lougenecker et al., 1940; Burns et al., 1963; Filipov, 1964).

Experiments with guinea pigs exposed to DDT and lindane while on diets with various percentages of ascorbic acid have clearly indicated a dependency on ascorbic acid for the enzyme induction to occur in the presence of organochlorine pesticides (Chadwick et al., 1973; Wagstaff and Street, 1971). Furthermore, insecticide residues were significantly increased in tissues of guinea pigs which were fed a vitamin C–deficient diet as compared with the controls (Chadwick et al., 1971).

Street and Chadwick (1975) suggested that similar effects of an ascorbic acid deficiency may also occur in primates, including man. In fact, Chadwick et al. (1971) investigated the effects of DDT and ascorbic acid deficiency in liver microsomal enzyme activities and lindane metabolism in female squirrel monkeys. Although the data are of a preliminary nature, their results clearly suggest that the general effects in this primate species are similar to those observed in the guinea pig. Finally, Street and Chadwick (1975) suggested that maintaining enzyme inductive capacity requires greater amounts of ascorbic acid than prevention of scurvy, because a diet with as little as 25 ppm ascorbic acid can prevent scurvy in guinea pigs, but ascorbic acid levels of 50 ppm and even 200 ppm were not so effective as 2000 ppm of ascorbic acid as enzyme-inducing agents.

Gaseous Pollutants

Carbon Monoxide

Chronic exposure of guinea pigs to CO increased the rate of consumption of and requirement for vitamin C. The toxic effects of chronic CO poisoning were prevented by taking 40 mg of ascorbic acid daily (Nizhegorodov, 1962). These data seem to support the suggestion by Klenner (1955) to make vitamin C the treatment of choice for both acute and chronic CO poisoning. Other studies have revealed that ascorbic acid treatment protected guinea pigs against exposure to high concentrations of hydrochloric acid vapor and nitric oxide (Ungar and Bolgert, 1938).

Ozone and Sulfates

Inhaled ozone and sulfates are known to effect the release of histamine from lung tissue, thus causing inflammation of respiratory tissue (Charles and Menzel, 1975; Easton and Murphy, 1967). Such exposure may be expected to place additional stress on asthmatics. However, ascorbic acid has been found to detoxify histamine (Subramanian et al., 1973). Furthermore, research (*in vitro*, rats, guinea pigs) has indicated that ascorbic acid pretreatment is able to exert significant protection against ozone intoxication (Matzen, 1957; Mittler, 1958; Pagnotto and Epstein, 1969).

Radioactivity

Many researchers have reported that X-ray exposure reduces levels of ascorbic acid in the body of guinea pigs and rats (Kretzschmer and Ellis, 1947; Monier and Weiss, 1952; Hochman and Block-Frankenthal, 1953; Oster et al., 1953; Dolgova, 1962). Kalnins (1953) found the guinea pigs given 50 mg/day of ascorbic acid were significantly better protected against the harmful effects of radioactivity as compared with controls provided with 1 mg/day. It was suggested that large doses of ascorbic acid acted as a detoxicant for histaminelike bodies or leukotoxins formed in the irradiated tissue.

Carrie and Schnettler (1939) prevented X-ray–induced leukopenia (reduced number of white blood cells in the blood) by giving 200 mg/day of ascorbic acid to the patients. Clausen (1942) reported a similar finding with X-ray–treated stomach cancer patients given daily injections of 500 mg of ascorbic acid. Using injections of 50 mg/day of ascorbic acid, Wallace (1941) prevented many of the general toxic symptoms of radiation sickness but did not prevent the intestinal changes resulting from powerful pelvic X-ray treatments. Additionally, it has been shown that ascorbic acid offers a protective effect for various enzyme activity from the effects of radiation exposure (Genazzani and Miele, 1959). Shapiro et al. (1968) suggested the need to examine the possibility of ascorbic acid as a radiation-protective agent in animals.

Prevalence of Vitamin C Deficiency

Numerous surveys exist which present data on the intake of vitamin C by populations of various countries. However, unless a survey is relatively current, it should be viewed with some caution as a result of changing food habits and new fortification practices. In surveys made in the United States, including the recent National Nutrition Survey, borderline (or below) intakes of vitamin C were often found, varying up to 30 percent for infants, children, and adults, especially in lower income groups. Of particular concern is the fact that average daily intakes of vitamin C have been steadily decreasing in the United States since 1944–1945 (Bogert et al., 1973).

OZONE TOXICITY AS AFFECTED BY DIETARY VITAMIN E AND SELENIUM

Vitamin E deficiency has been found to increase the susceptibility of rats to ozone toxicity, while vitamin E supplementation decreases ozone toxicity (Shakman, 1974; Goldstein et al., 1970; Roehm et al., 1971). Rat studies have indicated the protective role of supplemental vitamin E at levels of ozone that are approached or exceeded in southern California. For example, rats supplemented with vitamin E lived more than twice as long as rats deficient in vitamin E when both were exposed to 1.0 ppm daily. Rats deficient in vitamin E show the same pattern of pulmonary fatty acid alteration as rats exposed to ozone (Roehm et al., 1972). It has also been suggested by Roehm et al. (1972) that the molecular effects of ozone or inadequate vitamin E levels are identical.

In addition to vitamin E, there are several other naturally occurring dietary factors that reduce ozone toxicity. These substances, which are said to have antioxidant effects, are vitamin C, selenium, and sulfur-containing amino acids (Shakman, 1974).

Considerable attention has been focused on the role of vitamin E and antioxidants in preventing the peroxidation of highly unsaturated lipids. Vitamin E is capable of preventing the oxidation of the red blood cell membrane when it is exposed to H_2O_2 (Goldstein et al., 1970; Roehm et al., 1971, 1971a; Rose and Gyorgy, 1950). The absence of vitamin E may lead to membrane lipid destruction and subsequent hemolysis. Increased susceptibility of red blood cells to hemolysis is the earliest indication of a vitamin E deficiency, as was first discovered in animals and is now recognized to apply to human infants (Mackenzie, 1954; Gordon et al., 1958) and adults (Horwitt, 1960). Patients with various malabsorption syndromes such as acanthocytosis (Menzel, 1968) often develop hemolytic anemia, especially after ingesting "oxidant" drugs. Their enhanced susceptibility has been shown to be related to vitamin E deficiency.

In cystic fibrotic (CF) patients, in addition to pulmonary dysfunction, there is an apparent deficiency in essential fatty acids and a marked reduction in serum vitamin E, a fat-soluble vitamin. The observed deficiency in essential fatty acids may be due to a reduced ability of the patient to absorb these acids from the diet. This, in turn, may result in defects in membrane structure or stability. The reduced levels of vitamin E may be due to a lower requirement for this vitamin because of lower levels of fatty acids (Rosenlund et al., 1974). Because of the reduced ability of CF patients to absorb these acids, a recommended

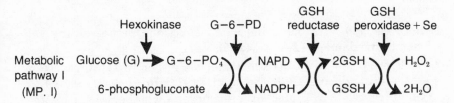

Figure 20. Mechanism by which reduced glutathione (GSH) helps maintain membrane stability following "oxidant" stress.

supplement of vitamin E may be of little value in protecting individuals exposed to ozone from defects in membrane structure or stability. Disruption of membrane stability in red blood cells may lead to hemolytic anemia.

Roles of selenium (Se), vitamin E (v. E), and sulfur-containing amino acids (S-aa) in maintaining cell membrane stability during oxidant stress, with particular reference to red blood cells (RBC), can be seen from the following explanation and Figures 20 and 21.

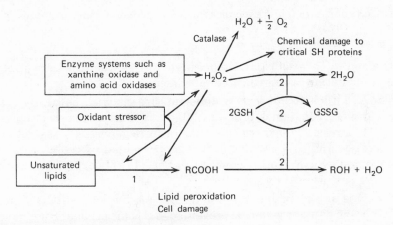

Vitamin E "blocks" reaction 1
Se, or a component of GSH–P_x, catalyzes reaction 2
a. Selenium as a component of GSH–P_x catalyzes the destruction of H_2O_2, lipid hydroperoxides (ROOH), and other hydroperoxides.
b. Vitamin E decreases the formation of hydroperoxides, particularly the lipid hydroperoxides.

Figure 21. Postulated functions of Se and vitamin E. *Source.* Hoekstra, W. G. (1974). Biochemical role of selenium. In: *Trace Element Metabolism in Animals*, Vol. 2. University Park Press, Baltimore, Maryland, p. 71. © 1974 University Park Press, Baltimore.

Known

1. Both Se and v. E have important functions in preventing cell membrane damage (Tappel, 1965).
2. V. E protected RBC against hemolysis either in the presence or absence of glucose (Rotruck et al., 1971; Rotruck et al., 1972). Significance: V. E protects the membrane whether MP. I is functional or not.
3. Only in the presence of glucose did dietary Se protect the cell membrane and hemoglobin against oxidative injury (Hoekstra, 1974). Significance: Se needs a functional MP. I to perform its antioxidant function.
4. Se is not involved with regeneration of GSH (Rotruck et al., 1972).
5. Dietary Se significantly enhances the ability of GSH to protect hemoglobin from oxidation, even in the presence of H_2O_2 (Roehm et al., 1971).
6. Selenium is thought to form a complex with GSH peroxidase (GSH-P_x). This complex is responsible for converting H_2O_2 molecules to water (Roehm et al., 1971).

Predictions of Biological Responses Based on Model Shown in Figure 21 (Hoekstra, 1974):

1. If Se and v. E are deficient, tissues high in H_2O_2 would be susceptible to peroxidation of unsaturated fatty acids.
2. If Se is sufficient but v. E is deficient, organs with naturally low levels of GSH-P_x (chick brain, rat placenta) would be susceptible to oxidant stress. Thus, there may occur brain degeneration, resorption of fetuses, and so on, which cannot be prevented by Se but can be prevented by v. E. *Note:* Reduced litter size in rats exposed to ozone during gestation (Veninga, 1968).
3. If v. E is sufficient but Se is deficient, production of ROOH and membrane damage would be prevented, but some organs do not have sufficient capacity to destroy the H_2O_2 produced because they may be low in catalase. Thus, the H_2O_2 would damage some critical SH proteins and lead to degeneration. *Note:* Potential health significance for acatalasemic individuals with a dietary deficiency in Se.
4. Species with higher liver H_2O_2 production are very susceptible to liver degeneration from the double deficiency.

5. The mechanism shows the sparing effect of v. E on the Se requirement. If v. E is in rather short supply, more ROOH is produced which requires a substantial quantity of GSH-P$_x$, thus requiring more Se and vice versa.

6. The sparing effect of S-amino acids on v. E and Se requirements may be explained by the fact that cysteine is a precursor of GSH and this represents a source of GSH other than that produced from reduction of GSSG.

The action of atmospheric oxidant (e.g., ozone) on the red blood cell is described in detail in the section on G-6-PD–deficient individuals.

In a study of 197 adults in Rochester, N.Y., it has been determined that the average serum vitamin E concentrations were 1.05 mg/100 ml. Of concern was the percentage of low values. Although no definitive evidence is available which would establish an undesirable level from the clinical standpoint, several researchers feel that on the basis of the *in vitro* hemolysis of RBC by peroxide, a vitamin E level below 0.5 mg/100 ml is indicative of a deficiency (Harris et al., 1961). Seven percent of this group was below the 0.5 mg/100 ml. If 7 percent of the United States population were considered deficient in vitamin E, this would involve at least 15 million people.

With regard to levels of selenium in the diet, there are no reports of Se deficiency in man (Allaway et al., 1968). Levels of Se in 210 whole bloods from donors in 19 cities and towns in the United States have been published, with extreme values ranging from 10 to 34 μg/100 ml with the mean value of 20.6 μg/100 ml. No blood was found deficient in Se. Total daily intake of Se has been estimated at 60 to 150 μg/day from food, with less than 1 μg/day coming from both water and air (Schroeder et al., 1970). Considering the widely separated sources of food today, it is not too likely that human deficiency exists (Schroeder et al., 1970).

However, natural Se deficiency has been reported in birds and animals in numerous countries of the world. In newborn animals, Se deficiency takes the form of a degenerative disease of the striated muscle; also, necrosis of the liver, retarded growth, and infertility may occur in animals deficient in Se (Schroeder et al., 1970).

Deficiency in both vitamin E and Se is considered here as a dietary condition which may predispose an individual to the effects of ozone toxicity. It should, however, also be pointed out that excessive amounts of vitamin E (Piedrabuena, 1970; Vogelsang et al., 1947; Greenblatt, 1957; King, 1949; Beckman, 1955; Ostwald and Briggs, 1966) and Se (Schroeder et al., 1970) have toxic effects in humans.

LEAD TOXICITY AND DIETARY DEFICIENCY

The toxic effects of Pb are very often found in young children, with a peak incidence at 1 to 3 years of age (Mahaffey-Six and Goyer, 1972). Susceptibility to Pb toxicity is known to be influenced by several physiological, environmental, and dietary factors, including (1) age (Mahaffey-Six and Goyer, 1972; Mahaffey, 1974), (2) season of the year (Sobel et al., 1938), (3) calcium (Ca) (Mahaffey-Six and Goyer, 1972; Mahaffey, 1974; Six and Goyer, 1970), iron (Fe) (Mahaffey-Six and Goyer, 1972), and phosphorus (P) dietary levels (Sobel and Burger, 1955), (4) vitamins A and D dietary levels (Sobel et al., 1938), (5) vitamin C (Holmes et al., 1939), (6) dietary protein (Baernstein and Grand, 1942), and (7) alcohol (Cramer, 1966).

Low dietary ingestion of Ca or Fe (20 percent of recommended levels) by rats significantly increased their predisposition to lead toxicity. For example, when rats were fed Ca at 20 percent of the recommended intake, 12 μg Pb/ml drinking water produced that same degree of toxicity as did 200 μg Pb/ml with a normal Ca diet (Mahaffey, 1974).

The concentration of Pb in tissues such as kidney and bone is much higher when rats are fed on a low Ca diet. However, the increase of Pb is much greater in kidney than in bone tissue. Decreasing the Ca concentration of the diet has also been found to increase susceptibility to Pb toxicity in the dog, horse, and pig (Mahaffey, 1974).

Iron deficiency combined with Pb exposure acts synergistically to impair heme synthesis. The Fe deficiency resulted in increased Pb content of kidney and bone. Although the quantity of Pb in the bones of animals with either a Ca or Fe deficiency is similar, the amount of Pb in the kidney is far higher in Ca-deficient animals than in Fe-deficient animals (Mahaffey-Six and Goyer, 1972) (see section on vitamin C and Pb).

The studies summarized here used animal models. The degree to which these dietary factors affect Pb toxicity in humans is not known. However, it is known that many young children have high exposure to Pb concomitant with low dietary intakes of Fe and Ca. Table 12 indicates the cumulative percentages of children not consuming quarter, half, and three-quarters of the recommended dietary intakes of Ca and Fe. The data were derived from children 2 to 3 years old, examined between 1968 and 1970 in the Ten State Nutrition Survey (Mahaffey, 1974; U.S. Dept. HEW, 1972). It can be seen that Fe deficiency occurs much more frequently than Ca deficiency; however, based on this and

Table 12. Cumulative Percentage of Distribution of Calcium and Iron Ingestion for Children 2 to 3 Years of Age from Low-Income Ratio States[a]

Calcium		Iron	
Ingestion (mg/day)	Percentage of Children	Ingestion (mg/day)	Percentage of Children
200	13.7	3.9	24.1
400	28.9	7.9	67.3
600	50.1	11.9	90.9
800[b]	65.9	15.0[b]	98.0

[a] Data derived from the Ten State Nutrition Survey, 1968–1970 (Ten State Nutrition Survey, 1972).

[b] Recommended dietary allowances for children 2 to 3 years old, 1968 (NAS, 1968).

Source. Mahaffey (May, 1974).

other dietary surveys, Ca ingestion of less than quarter the recommended dietary intake occurs in 5 to 15 percent of children from low-income groups.

Additionally, rat studies have shown that the Pb concentration in the blood and the amount of Pb present in the bones was approximately two times higher in vitamin D–fed animals as compared with controls (Sobel et al., 1938). Such evidence seems to explain previous observations of increased severity of Pb poisoning under the influence of vitamin D or ultraviolet rays. Consequently, the effect of vitamin D on Pb metabolism seems to be similar to its action on Ca (Six and Goyer, 1970).

Of considerable concern is the possible effect of a milk diet on lead metabolism. A European study has indicated that a milk diet causes dramatic increases (33 to 57 times) in Pb absorption, principally due to greatly increased absorption of Pb via the intestine. It should be noted that cow's milk has appreciable quantities of Pb (Mitchell and Aldous, 1974). Such enhancement of Pb absorption is thought to be related to low values of Fe content in the milk as well as the presence of lactose and other undetermined factors which may facilitate the absorption of Pb (Kello and Kostial, 1973).

MAGNESIUM DEFICIENCY AS A FACTOR IN FLUORIDE TOXICITY

Information derived from research conducted on several animal species has indicated that dietary levels of Mg affect the toxicity of fluorides. For example, when very high levels of Mg and fluoride were fed to growing chicks, it caused noticeable leg weakness (Gardner et al., 1961) and reduced the mineral content of bone but increased the Mg content of bone. According to Marier (1968), commenting on Griffith et al. (1964), this condition of the bone implies a rachiticlike state which may be caused by a high-Mg, low-citrate condition plus the presence of fluoride. Belanger et al. (1958) have noted that high fluoride supplements can produce a rachiticlike condition. Also, rachitic bone has an increased Mg content (Fourman and Morgan, 1962).

Animal studies with dogs and rats (Chiemchaisri and Philips, 1963) revealed that, in Mg-deficient dogs, fluoride supplements prevented soft-tissue calcification but not the muscle weakness and convulsions. In rats, the dietary fluoride, although having no protective effect on soft tissue calcification, aggravated the hypomagnesemia condition which intensified the occurrence of convulsive seizures.

According to Marier et al. (1963), the symptoms of Mg deficiency are quite similar to those of fluoride intoxication. This is most probably due to a fluoride-induced increase in the uptake of Mg from plasma into bone. Marier (1968) notes that even small increases in bone Mg may be serious, since bone can contain up to 63 percent of body Mg, whereas body fluids account for only about 1.0 percent. Table 13 indicates some of the symptoms common to both Mg deficiency and fluoride intoxication.

Since Mg deficiency may influence the toxicity of fluoride, it is important to determine the dietary requirement for Mg. According to Seelig (1964), the assumption that the average daily intake of Mg is enough to maintain equilibrium in the normal adult is highly suspect. An analysis of metabolic data shows that the minimal dietary requirement is not 220 to 300 mg/day or even 5 mg/kg/day, as has also been suggested, but probably at least 6 mg/kg/day.

Seelig (1964) reports that clinical metabolic data indicate that with intakes below 6 mg/kg/day a negative Mg balance is likely to develop, especially in men. High protein, calcium, vitamin D, and alcohol ingestion function either to prevent retention or to increase the requirement of Mg, especially in individuals on low Mg intakes. The Western diet is calculated to supply an average of 250 to 300 mg of Mg daily, which is less than 5 mg/kg/day. Since the Western diet is usually rich in protein, Ca, vitamin D, and alcohol, Seelig (1964) suggests that

Table 13. Symptoms Common to Both Fluoride Intoxication and Magnesium Deficiency

1. Leg cramps
2. Muscular twitching
3. Tetaniform convulsions (with normal serum Ca)
4. Two- to threefold increase in serum P at time of convulsions
5. Optical neuritis
6. Bone exostoses or soft tissue calcification

Note. With the exception of symptom 6, all the above conditions were noted in human subjects.

Source. Marier (1968). The importance of dietary magnesium with particular reference to humans. National Research Council of Canada, No. 10, 173. (From *Z. Vitalstoffe-Zivilisationskrankheiten* **13:** 144–149, 1968, and reproduced with permission of the author.)

the optimal daily intake of magnesium should be between 7 and 10 mg/kg/day.

Although the data suggest that individuals living in the United States have less than sufficient Mg in the diet as well as having daily exposure to fluoride (when exposed to fluoridated drinking water at 1.0 ppm) via food and water between 3.5 to 5.5 mg (Marier and Rose, 1966), the precise risk is difficult to ascertain because of the uncertainties of extrapolation from animal studies to man and the lack of supporting human epidemiologic data.

MANGANESE TOXICITY AS AFFECTED BY DIETARY IRON DEFICIENCY

The toxic effects of direct manganese (Mn) exposure are usually found only in occupational settings. In reported cases of Mn toxicity, miners have been exposed to Mn dusts for various lengths of time from about a year to greater than a decade. The frequency of industrial manganism is reportedly quite low. It is often characterized by a self-limited psychiatric disorder, at the end of which neurological symptoms appear. These symptoms may persist even after the excess Mn becomes cleared from the tissues (Mena et al., 1969).

Several studies have indicated that animals deficient in Fe tended significantly to increase their absorption of Mn from the gastrointestinal tract and to incorporate it into the porphyrin part of the hemoglobin molecule of the red blood cell (Borg and Cotzias, 1958, and Pollack et al., 1965). A more recent study involving 36 human subjects of various nutritional states revealed increased Mn absorption in the presence of Fe deficiency. For example, anemic individuals absorbed Mn via the intestine at a rate of 150 percent greater than "normal" individuals. The authors suggested that if an individual became anemic while being exposed to high levels of Mn, the person's tissue concentrations of the metal would quickly become gross. Consequently, it was concluded that workers with Fe deficiencies would be at high risk with respect to Mn toxicity (Mena et al., 1969).

Recent rat studies have strongly supported the previously discussed research by indicating that Mn concentrations in tissues of rats deficient in Fe and fed $MnCl_2$ for 15 days were higher than in tissues from the control group. For example, the respective concentrations of Mn in the liver and kidney were 2.5 and 2 times greater in the Fe-deficient animals as compared with the controls (Chandra and Tandon, 1973).

A further complication is that Mn is excreted from various organ systems at different rates. For example, the presence of Mn in the central nervous system (CNS) has the longest half-life as compared with Mn in any other system in the body (Dastur et al., 1971).

NUTRITIONAL INFLUENCES ON THE TOXICITY OF PESTICIDES

Inadequate dietary protein enhances the toxicity of most pesticides but decreases or fails to affect the toxicity of a few. Considerable data using rat models have led to the following conclusions (Shakman, 1974):

1. At one-third normal dietary protein, toxicity is slightly increased for most pesticides.
2. At one-seventh normal dietary protein, carbonate carbaryl, the organophosphate parathion, and the phthalidimide captan significantly increase toxicity.
3. At one-seventh or less normal dietary protein, the toxicity of heptachlor is decreased and that of dimethoate is unaltered.

The type of dietary protein may also affect the pesticide toxicity (Boyd, 1969, and Shakman, 1974).

Numerous factors may offer a partial explanation concerning the variation in pesticide toxicity in relation to dietary factors. These factors include liver microsomal enzyme induction, quantity of body fat, and levels of methionine, carbohydrate, fat, riboflavin, and nicotinic acid in the diet (Shakman, 1974).

Enzyme Induction

Organochlorine insecticides are well-known enzyme inducers. However, some organophosphorous compounds are thought to inhibit drug-metabolizing enzymes (Durham, 1967). Some insecticides (heptachlor and dimethoate) must be structurally altered in the body (liver) in order to become toxic. Decreased dietary protein decreases drug-metabolizing capacity by reducing enzyme concentrations. A protein deficiency which affects a reduction in microsomal enzyme activity reduces production of toxic metabolites (Weatherholtz et al., 1969).

Methionine

Rats reared on a methionine-deficient diet and exposed to DDT experienced depressed growth and decreased vitamin A stores. The addition of methionine to the diet eliminates the toxic effects of DDT (Shakman, 1974). It has been suggested that methionine influences the absorption of vitamin A from the intestinal mucosa into the lymphatics via chylomicron formation; DDT may inhibit the interaction of vitamin A ester with the chylomicrons (Tinsley, 1969) (see "Vitamin A Deficiency").

Carbohydrates

Digestion and absorption of amino acids are reduced by a high proportion of nonusable carbohydrates in food in the diet. The effect of carbohydrates on pesticide toxicity is therefore considered an indirect one, that is, causing a modification in protein availability (Shakman, 1974).

Fat

Lack of food leads to the transfer of DDT from mammalian body fat deposits to the blood and may produce concentrations potentially toxic to the central nervous system. It has been found that "overweight" mammals, fish, and birds are more resistant to DDT poisoning than thinner controls (Keane et al., 1961).

Deficiencies in the levels of essential fatty acids, riboflavin,* and nicotinic acid enhance the toxicity of dieldrin (Shakman, 1974; Durham, 1967).

SELENIUM: INTERACTIONS WITH CADMIUM AND MERCURY

The administration of Se compounds provides complete protection against the known toxic effects of Cd on reproductive processes such as the selective damage caused by Cd in the nonovulating ovaries (Kar et al., 1959; Parizek et al., 1968), in the placenta (Parizek et al., 1968a), and toxicity during the last stages of pregnancy (Parizek, 1965). Selenium compounds have also been demonstrated to prevent the teratogenic effects of Cd (Holmberg and Ferm, 1969) and to decrease the mortality of experimental animals administered doses of Cd which were lethal in corresponding control groups (Gunn et al., 1968).

Animals administered a high dose of mercuric compounds which were lethal to controls survived when concomitantly given compounds of Se. Following Se treatment, the experimental animals exhibited an increased retention of Hg and changes in Hg distribution in the body (Parizek et al., 1971). Further studies have revealed a decreased passage of Hg into fetuses and into milk from mothers given Se compounds. It should be pointed out that high Hg levels in tuna are accompanied by a high Se content and that Hg in such cases is significantly less toxic (Ganther et al., 1972; Potter and Matrone, 1973). (See section on O_3 and vitamin E for a consideration of how Se affects O_3 toxicity, p. 102.)

ZINC DEFICIENCY AND CADMIUM TOXICITY

Zinc (Zn) plays an important role in the maintenance of human health. It is a component of various enzyme systems, including carbonic

---

* Morgan (1959) and Bogert et al. (1973) have indicated that riboflavin deficiency is quite prevalent in the United States. Specifically, 30 percent of women and 10 percent of men ages 30 to 60 ingest less than two-thirds of the RDA.

anhydrase and alkaline phosphatase. A nutritional deficiency of Zn in animals causes impaired growth, testicular atrophy, and fibrotic changes in the esophagus, while excessive levels of Zn are associated with a wide variety of toxic symptoms. However, it has been suggested that the proposed Zn toxicity is actually caused by small quantities of Cd which are often present along with Zn, since the symptoms are nearly identical with Cd toxicity (e.g., damage to kidneys and gastrointestinal tract in mammals, renal tubular injury, proteinuria, anemia, and hypertension) (Shakman, 1974).

Cadmium is known to be a metabolic antagonist of Zn (Jacobs et al., 1969). Schroeder (1967) has reported that Cd competes with Zn for thiol group binding sites on the protein metallothionein. In rats, injections of Cd induce hypertension which can be cured by replacing some of the Cd with Zn (Schroeder, 1967). Finally, Parizek (1957) noted that Zn provided a protective effect against testicular necrosis caused by Cd.

TRACE ELEMENTS AND CANCER

A deficiency in dietary iron has been closely associated with the development of the Plummer-Vinson syndrome. This syndrome is most often found in northern Scandinavia and is characterized by a high risk of cancer developing in certain areas of the upper alimentary tract such as the pharyngeal and esophageal mucosa. The frequency of the Plummer-Vinson syndrome has shown a significant reduction in Sweden since the initiation of supplementation of flour with iron and other nutrients (Wynder et al., 1957; Chisolm, 1974; Wynder et al., 1975).

Wynder et al. (1975) have suggested that dietary deficiencies of both riboflavin (see p. 96) and iron may play a similar role in the promotion of cancer. They hypothesized that since AHH, a respiratory enzyme, has been implicated in the bioactivation of DMBA to a carcinogenic form and since respiratory enzymes such as AHH are somewhat dependent on the nutritional status, this altered state of enzymatic activity may enhance the bioactivating capabilities of AHH with respect to pre-carcinogens.

Other trace elements have been at least statistically associated with the development of cancers. For example, esophageal cancers are fairly frequent in parts of Africa where soil contains negligible amounts of molybdenum (Mo) (Burrell et al., 1966). Data supporting this suggestion are derived from retrospective epidemiological studies in the United

States which indicate a positive association between low levels of Mo in drinking water supplies and increased incidences of esophageal cancer (Berg et al., 1973). More research is certainly needed to examine the Mo deficiency–cancer association.

Several of the naturally occurring antioxidants such as Se and vitamin C as well as commercially prepared antioxidants, including butylated hydroxytoluene (BHT) and butylated hydroxyanisole (BHA), have been associated with lower cancer rates. For example, Se has been reported to reduce tumor frequency in experimental settings (Shamberger, 1970) as well as in retrospective epidemiologic associations (Shamberger and Willis, 1971). In the latter case, it was reported that cancer rates are reduced in areas in the United States where higher Se levels are found in plants, milk, and human blood. With regard to vitamin C, it has been suggested that low vitamin C ingestion is linked to areas of high stomach cancer rates (Berg, 1975; Dungal and Sigurjonsson, 1967). These results are in fundamental agreement with the research of Haenszel et al. (1972), Haenszel (1967), and Graham et al. (1972), which reported a negative statistical relationship between ingestion of raw vegetables and stomach cancer.

From the nature of the statistical associations mentioned above, it becomes very apparent that major research efforts are needed to clarify many of the suggested dietary relationships indicated here.

Diseases and Pollution Exposure

The presence of certain diseases is an important factor in the predisposition of certain individuals to the harmful effects of environmental pollutants.

HEART-LUNG DISEASE AND AIR POLLUTION

Pollution Episodes

The history of catastrophic pollution episodes [e.g., Meuse Valley, Belgium, 1930 (Firket, 1931); Donora, Pennsylvania, 1948 (Schrenk et al., 1949); and London, England, 1952 (Ministry of Health, 1954)] (Table 14) has clearly indicated that the very young and the very old were affected more severely than other age groups. Chronically ill individuals, especially those with cardiovascular (affecting the heart and blood vessels) and respiratory diseases, were the most seriously affected. The mortality rate was much higher in these groups than in any other. An overall perspective revealed that deaths from cardiovascular disease occurred early and fell off markedly and deaths from pulmonary diseases most often started to occur on the second or third day and continued over a longer time interval.

Table 14. Acute Air Pollution Episodes the World Has Experienced

Place	Time	Effect and Number	Conditions and Probable Cause
Meuse Valley, Belgium; coke ovens, blast furnaces, steel, glass, zinc, and sulfuric acid plants	Dec. 1–6, 1930	60 deaths, "thousands ill"; coughing, breathlessness, chest pain, eye and nose irritation experienced	Inversion, stagnation in 15-mile river valley for 1 week; smoke and irritant gases; sulfur oxide, sulfuric acid mist, and fluorides suspected; estimated sulfur dioxide 25–100 mg/m^3 (10–40 ppm)
Donora, Pa., U.S.; zinc smelter, wire coating mill, steel mill, sulfuric acid plant	Oct. 27–31, 1948	6000 of 14,000 population ill, 1400 sought medical care, 17 died; coughing, sore throat, chest constriction, burning and tearing eyes, vomiting, nausea, excessive nasal discharge	Temperature inversion and fog along horseshoe-shaped valley of Monongahela River; sulfur oxides, smoke, and zinc compound particulates present; sulfuric acid mists likely; estimated sulfur dioxide of 1.5–5.5 mg/m^3 (0.5–2 ppm)
Poza Rica, Mexico; petrochemical plant, hydrogen sulfide recovery system	4:45 A.M.–5:10 A.M., Nov. 24, 1950	22 deaths, 320 hospitalized; acute hydrogen sulfide poisoning, unconsciousness, vertigo, severe irritation of respiratory tract, loss of sense of smell	Low inversion layer, fog, weak winds; hydrogen sulfide released when burner on 4-day-old sulfur recovery plant failed under increased hydrogen sulfide flow rate; release lasted for only 25 min

London, England	Dec. 5–9, 1952	3500–4000 deaths in week of Dec. 5–12 in excess of expected norm of like weeks; causes of death, chronic bronchitis, bronchopneumonia, and heart disease; increased hospital admissions for respiratory and heart disease	"Pea soup" fog and temperature inversion covered most of the U.K.; smoke and sulfur dioxide accumulations in stagnated air; reported smoke highs of 4.5 mg/m^3 and sulfur oxide highs of 3.75 mg/m^3 (1.4 ppm)
London, England	January, 1956	1000 excess deaths charged to a pollution episode	Extended fog conditions similar to 1952 episode; resulted in Parliament's passing Clean Air Act
London, England	Dec. 5–7, 1962	700 excess deaths and increased illness charged to a pollution episode; emergency medical care plan functioned	Severe fog and inversion; sulfur dioxide levels higher than in 1952, but particulates were lower; alert system operated

Source. From *Environmental Protection* by E. T. Chanlett. Copyright 1973, McGraw-Hill Book Company. Used with permission of McGraw-Hill Book Company.

In the Donora episode, 60 percent of all the individuals over the age of 65 became ill, half of these severely ill. The most seriously affected individuals became ill on days 2 and 3 of the episode. An investigation of those affected indicated that many were immunologically hypersensitive to some of the pollutants. An overwhelming percentage (80 percent) of those with a medical history of heart disease or chronic bronchitis became ill, although fewer were among the severely ill as compared with the hypersensitive patients (Schrenk et al., 1949).

In the London episode, the largest excess of deaths was related to bronchitis. Eight times more people died from bronchitis as would normally be expected at that time of year, and three times the predicted "normal" number of deaths from pneumonia occurred. There also appeared to be a significant increase in death from myocardial degeneration and coronary heart disease (Ministry of Health, 1954).

In addition to data derived from acute exposure and the response of people with diseased states to pollution levels, there are considerable numbers of clinical epidemiologic studies dealing with lower than lethal levels of pollution and the effects on man's health. For example, in the New Orleans, Louisiana area, a dramatic increase in acute asthmatic attacks occurred in 1 week. Three hundred people entered Charity Hospital for treatment, and nine deaths occurred. A strong correlation was established between the high incidence of attacks in one part of the city and incomplete combustion of hemp and other materials with a silica content in a garbage dump in the same area. Immunological tests revealed that all those affected with acute asthmatic attacks responded positively to antigens prepared from the suspected material as compared with only 20 per cent of the controls (Lewis et al., 1962). In Nashville, Tennessee, the incidence of acute asthmatic attacks experienced by susceptible patients seemed to correlate directly with the measured variations in sulfate pollution levels (Zeidberg et al., 1966).

A relationship between the causes and development of bronchitis and high levels of sulfur dioxide has also been postulated in a number of studies dealing with frequency of bronchitis in certain groups of individuals. One such study considered London postmen working outside. Individuals in the study were matched for job similarity, social status, pay, and so on. In London, the prevailing SSW wind causes a large increase in smoke pollution over the central and NE areas. Consequently, in the NE part of London, absence from work because of bronchitis was nearly twice that of postmen in all the other quadrants of

London. The retirement and death rates were also much higher. Furthermore, a study comparing individuals living in two United States communities, one with high and the other with low pollution levels, indicated a generally lower level of pulmonary function in those residents of highly polluted areas (Fairbairn and Reid, 1958).

Another study considered the effects of air pollution on pulmonary function with particular emphasis on the chronically ill. In this study, patients with emphysema who were placed in a room with filtered air showed improvements in lung function. However, when ambient Los Angeles, California, air was pumped into the room, their lung function decreased. This difference did not happen to "normal" individuals experiencing the same conditions (Motley et al., 1959).

Besides increasing pulmonary resistance and causing attacks of asthma and asthmalike diseases, it has been strongly suspected that air pollution may influence the onset of other acute respiratory illness in any exposed person and severely potentiate infection in those who are chronically ill (Carnow, 1966). The percentage of adult individuals in the general population with such cardiopulmonary diseases includes asthmatics, 2 to 5 percent; persistent chronic respiratory disease symptoms, 3 to 5 percent; heart disease severe enough to limit activity, 7 percent (Finklea et al., 1974).

Estimations of Increased Susceptibility of High-Risk Groups to Respiratory Irritants: SO_2, Particulate Sulfates (PS), and Particulates (P)

An exciting attempt to quantify the risks of individuals who are considered hypersusceptible to the effects of respiratory tract irritants has been developed by Finklea et al. (1974). With regard to SO_2, PS, and P, the authors indicated that aggravation of preexisting cardiorespiratory symptoms in the aged, as well as aggravation of asthma and irritation of the respiratory tract (Carnow, 1970; Carnow and Carnow, 1974; and CHESS, 1974), appeared to occur at lower levels than the present ambient air quality standards for SO_2 and P. It was suggested that the apparent "built-in" safety margins in the air standards are quite minimal; in all cases, they are less than the standard itself. However, the irritant effects recorded at SO_2 and P levels below the present standards have been considered to be possibly due to the action of acid sulfate aerosols which have their origins from the interaction of SO_2, P, and aerosols in the atmosphere. Table 15

Table 15. Ambient Air Quality Standards for SO_2, TSP, and Particulate Sulfates and Safety Margins for High-Risk Groups

High-Risk Group	Effect	24-Hr Thresholds ($\mu g/m^3$)			Annual Thresholds ($\mu g/m^3$)		
		Sulfur Dioxide (SO_2)	Total Suspended Particulate (TSP)	Particulate Sulfates (PS)	SO_2	TSP	PS
Children	Aggravation of respiratory symptoms	ID (Inadequate data)	ID	ID	200	100	11
Elderly	Aggravation of respiratory symptoms	300–400	250–300	ID	ID	ID	ID
Asthmatics	Aggravation	180–250	100	8–10	ID	ID	ID
Bronchitics	Increased prevalence	ID	ID	ID	95	100	14
Present standard		365	170	No standard	80	75	No standard

Source. Adapted from Finklea et al. (1974).

indicates a "best judgement" relationship of pollutant (SO_2, TSP, PS) levels and the health responses of various high-risk groups according to Finklea et al. (1974).

However, it should be emphasized that the estimates in Table 15 are tentative conclusions. The whole area of SO_2, PS, and P and their individual and collective health effects is broadly recognized by most leading authorities, including Finklea et al. (1974), as presenting enormous difficulties in the attempt to distinguish the effects of the specific pollutants. Analyses of the health effects of oxides of sulfur by Schimmel and Murawski (1976), Battigelli and Gamble (1976), and Tabershaw (1976) likewise point out the controversy centering on the toxicity of acid sulfates and their relationship to SO_2 levels. Certainly, more information is needed to permit more confident conclusions from which policy decisions can evolve; this is especially true with regard to such technological innovations as the catalytic converter. The importance of resolving the controversy surrounding the toxicity of SO_2 as compared with acid and particulate sulfates in the role of standard setting becomes especially acute when the magnitude of the affected high-risk groups is realized.

Waldbott (1973) has suggested that certain people may be allergic (hypersensitive) to sulfur dioxides because asthma, chronic bronchitis, and eye and nasal symptoms have long been associated with levels of sulfur oxides. Recently, Charles and Menzel (1975) and Menzel (1976) have proposed an explanation for the extreme hypersensitivity of asthmatics to oxides of sulfur. They found that ammonium sulfate ions release histamine from guinea pig and rat lung tissues as well as decreasing the rat tidal volume. These data have been interpreted by Menzel (1976) as meaning that most of the bronchoconstriction associated with oxides of sulfur exposure is caused by histamine release. Calculations based on research with the rat lung have indicated that the rate of normal sulfate clearance from the lung is not sufficient to eliminate all the sulfate inhaled at ambient levels (20 $\mu g/m^3$). It was suggested that the sulfate salts which remain in the lung may be sufficient to result in a significant degranulation of mast cells with subsequent release of histamine.

Respiratory Irritants and Sudden Infant Death Syndrome

A particularly important high-risk group with regard to respiratory tract pollutants includes infants less than 1 year of age. The sudden infant death syndrome (SIDS) results in 30 to 40 percent of all deaths between the first and twelfth months of life (Greenberg et al., 1973). Despite its widespread occurrence, the etiology of this disease remains to be established. Over the years, various theories have been presented, only to fall considerably short of providing an adequate explanation (Valdes-Dapena, 1967). It is quite possible that the SIDS may be influenced by multifactorial agents. Certainly this is true of numerous diseases, including hypertension and kidney and heart disease. Consistent relationships of SIDS with numerous factors, including poverty, winter season, colonization or infection with different respiratory viruses, and history of mild upper respiratory tract symptoms, have been reported (Valdes-Dapena, 1967). Furthermore, research by Ehrlich (1966) with animals indicated that NO_2 exposure enhances susceptibility to bacterial aerosols. As a result of a variety of factors, including (1) the frequency of SIDS increases in the winter, when SO_2 and several other pollutant concentrations are highest, (2) the previous association of respiratory tract symptoms with SIDS, (3) SO_2 being a major respiratory irritant, and (4) the previously cited research of Ehrlich (1966), Greenberg et al. (1973) investigated the relationship of environmental factors, including SO_2 levels, on the incidence of SIDS in Chicago. They were unable to find any independent relationship between SO_2 levels and SIDS, however. This is certainly an area where considerable research effort needs to be directed, especially with regard to other pollutants including CO and NO_2.

Photochemical Oxidants

Ozone, like NO_2 and SO_2, acts directly on the lining of the respiratory tract. Ozone is known to affect the functioning of pulmonary macrophages adversely and to induce thinning of the walls of small pulmonary arteries (P'an et al., 1972). Such toxic actions by ozone are clearly indicative of the development of chronic pulmonary disease and early emphysema. Ozone also is known to disrupt the functioning of several important enzymes (alkaline phosphatase and 5-ribonucleotide phosphohydrolase) in lung tissue (Hathaway and Terrill, 1962). Table 16 summarizes the major health effects of ozone on humans according to concentration and duration of exposure. Based on the evidence presented in Table 16, it is quite clear that those individuals at high risk to the respiratory-irritating effects of ozone include young children, individuals with chronic pulmonary disease, and asthmatics as well as individuals engaged in vigorous physical activity (see pp. 139 to 143 for discussions of the toxic effects of CO and NO_2).

CARDIOVASCULAR DISEASES AS AFFECTED BY DRINKING WATER QUALITY

Water Mineralization: Hard Water versus Soft Water

There is considerable controversy which surrounds the general subject of human health and drinking water quality. One of the most researched and discussed theories associated with the health effects of drinking water is that the incidence of cardiovascular disease (CVD) is inversely related to the hardness of the water. Thus, it has been variously proposed that people living in hard-water areas tend to be protected from CVD, and those people living in soft-water areas are more susceptible to the development of CVD.

The initial association of soft drinking water and CVD was presented in Japan by Kobayashi (1957), who linked acidity in river water with the incidence of cerebral hemorrhage. It was found that areas in Japan with more acidic rivers had a higher incidence of mortality from this disease.

Subsequently, Schroeder (1960) and Morris et al. (1961) demonstrated a highly significant inverse correlation between CVD mortality and water hardness in the United States and England and Wales, respectively. Additional research by Muss (1962) indicated that CVD mortality for 1931–1940 of a section of New York City also inversely related to water hardness. More specifically, the hard-water section had 10 percent lower CVD mortality than the soft-water section.

Several studies in the middle and later years of the 1960s supported the previously mentioned inverse-relationship theory. For example, Biersteken (1967) noted a statistically significant negative correlation between water hardness and CVD in women but not in men in the Netherlands, and Anderson et al. (1969) indicated that soft-water areas of Ontario, Canada, had increased coronary heart disease death rates as compared with hard-water areas for combined male and female rates.

In an English study, CVD mortality was examined in a community where the water had been artificially softened. It should be noted that, in all previously mentioned studies, soft water was naturally occurring. In a report of this study, Robertson (1968) indicated that mortality from CVD significantly increased after the community decreased the hardness of its water (488 ppm to 100 ppm total hardness). Winton and McCabe (1970), commenting on a personal communication from Robertson, indicated that following the evidence of the increased CVD, the water hardness was subsequently increased to 225 ppm. It was

Table 16. Relationship of Ozone and Photochemical Oxidant Exposure to Human Health Effects and Recommended Alert and Warning System Levels

Recommended Episode Levels	Ozone (ppm) and Photochemical Oxidants	Duration of Exposure	Health Effects
	0.70	2.0 hr	Soreness of upper respiratory tract; tendency to cough while taking deep breaths; significant increase in breathing difficulty. These conditions were made worse by 15 min of light exercise.
Emergency	0.50[a]	2.75 hr	Measurable biochemical changes in blood sera enzyme levels and in red blood cell membrane integrity; some subjects became physically ill and unable to perform normal jobs for several hours.
	0.37	2.0 hr	Impairment of pulmonary function in young adults probably due to a decreased lung elastic recoil, increased airway resistance, and small airway obstruction.
	0.37[a]	2.75 hr	Significant biochemical changes in blood sera enzyme levels and in red blood cell membrane integrity, but less severe than at 0.50 ppm; some subjects became physically ill and unable to perform normal jobs for several hours.
Red alert	0.30	—	Precipitous increase in rates of cough and chest discomfort in young adults.
	0.25	—	Greater number of asthma attacks in patients on days when daily maxima equalled or exceeded 0.25 ppm during a 14-week period.
	0.25[a]	2.75 hr	Biochemical changes in blood sera enzyme levels.
Yellow alert	0.10	1 hr	Breathing impaired.
	0.10	—	Tokyo elementary school children had significantly reduced respiratory function associated with ozone levels less than 0.1 ppm during a long-term epidemiologic study. Beginning of headache without fever in young adults; median age 18.6 years.

Table 16 (Continued)

Watch	0.07	2-hr average	
	0.065	—	Impairment of performance of student athletes during running competition.
	0.05	15–30 min	Threshold of respiratory irritation.
	0.02	—	Odor perception.
	0.005	—	Decreased electrical activity of the brain.

[a] These values were derived by standard EPA analytic procedures. They exceed the "absolute" value of ozone determined by the ultraviolet photometer method by approximately 25 percent. Thus these stated values may be approximately 25 percent lower than reported here. For example, 0.50 ppm would be approximately 0.40 ppm; 0.37 ppm would be approximately 0.30 ppm; and 0.25 ppm would be approximately 0.19 ppm.

Source. Health Effects and Recommended Alert and Warning System for Ozone. Environmental Health Resource Center, Illinois Institute for Environmental Quality, 1975.

suggested that this additional hardness would result in 30 fewer deaths per year in the community's middle-aged men.

At this point, it is important to consider whether the lower heart disease death rates may be the result of a "beneficial factor" in the hard water or whether the higher heart disease death rate could be caused by a "detrimental factor" in the soft water.

Research into this aspect of the problem has revealed some interesting potential clues to the inverse-relationship phenomenon. Certainly, it is well known that individuals with high levels of blood cholesterol and other lipids are more likely to develop coronary heart disease (Keys, 1975). Yacowitz et al. (1965) demonstrated that levels of blood cholesterol and other lipids could be lowered in men and women by doubling the calcium intake. In a related study, Vitale et al. (1957) demonstrated that magnesium helps to prevent lipid deposition in the arteries of rats. However, further research examining the association of magnesium with lipid deposits on the arteries of humans (see Jankelson et al., 1959, and Brown et al., 1958) has indicated no correlation between magnesium and blood lipid levels. Yet, a study by Goldsmith and Goldsmith (1966) implied that the elevated levels of magnesium present in hard water may have a positive effect against coronary heart disease by preventing blood clot formation. The authors reported the presence of high serum magnesium levels in hibernating animals. Since the blood of hibernating animals is known to be naturally prevented from clotting, this association seems to support or, at least, not contradict the suggested role of magnesium in reducing the incidence of coronary heart disease. Furthermore, the direct addition of magnesium to the blood of animals, *in vitro* and *in vivo*, inhibits blood clotting. Finally, Goldsmith and Goldsmith (1966) noted that women who take birth control pills have a lower serum magnesium level and faster clotting time.

Although the above-cited references support the contention that substances, namely, calcium or magnesium, in hard water may help to prevent coronary heart disease, Schroeder (1974) has reported that soft water is corrosive and can be demonstrated to leach cadmium from galvanized pipes. Schroeder (1965, 1974) indicated further that cadmium is a direct causative agent of hypertension in rats. This cadmium hypertension in rats is quite similar to human hypertension with regard to elevated blood pressure, increased mortality, renal arteriolar sclerosis, enlarged hearts, and increase in the severity of atherosclerosis. Although human studies designed to verify the rat studies of Schroeder are needed, retrospective mortality studies have revealed that air levels of cadmium were highly correlated with deaths from hypertensive heart disease in 28 urban areas of the United States (Carroll, 1969).

Elevated levels of sodium in drinking water have been implicated as a possible agent in the development of certain types of CVD (Wolf, 1976). Other studies have correlated inverse relationships of calcium and sodium levels in drinking water (Wolf and Moore, 1973). Previous research (see Dahl and Love, 1954, 1957; Dahl et al., 1972) has implicated dietary salt consumption as a possible causative agent in the development of hypertension (Figure 22). However, since the normal dietary intake of salt (NaCl) varies between 4 and 10 g/day and accounts for greater than 90 percent of the usual sodium exposure (Schroeder, 1974), it seems difficult to understand how sodium from drinking water may play a significant role in the development of hypertension in normal individuals. However, individuals at increased risk (i.e., people with confirmed or incipient congestive heart disease, hypertension, renal disease, or cirrhosis of the liver) to the development of CVD, who must also be on a low-sodium diet could be critically affected by elevated sodium levels in drinking water. For example, Furstenberg et al. (1941) have reported that high levels of sodium in the drinking water resulted in the initial failure of a low-salt diet to treat Ménière's disease

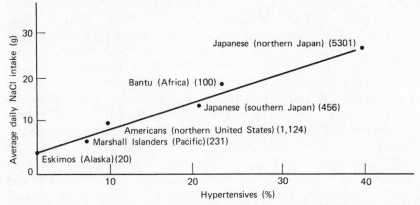

Figure 22. Comparison of the prevalence of hypertension among populations according to their average salt (NaCl) intake. Number of persons studied given in parentheses. *Source.* Weinsier (1976).

effectively. According to Elliot and Alexander (1961), the health of patients on sodium-restricted diets may be adversely affected by large amounts of sodium in the drinking water. They reported several cases of recurrent episodes of heart failure at home which ceased after substitution of a low-sodium drinking water source for a high-sodium source was made. Russell (1969) has indicated that patients following a restricted sodium diet (500 mg Na/day) and consuming 2.5 liters of water/day containing 125 mg/l would consume 64 percent of the daily sodium allowance from the drinking water. Thus, high levels of sodium in drinking water may be an important contributing factor in the development of a variety of heart disorders in high-risk patients.

The occurrence of elevated sodium levels in drinking water of high-risk individuals is an environmental pollutant problem which has recently become recognized in the eastern part of Massachusetts (Huling and Hollocher, 1972). Since the mid-1950s, there has been a profound increase in the use of salt, principally NaCl, to melt ice and snow from Massachusetts state highways. In fact, the rate of application of salt to state highways is about 20 metric tons of total salt (NaCl + $CaCl_2$) per lane mile per year. One of the environmental consequences of increased application of salt to the Massachusetts roadways has been the unintentional contamination of various community water supplies. For example, 15 communities in eastern Massachusetts have reported greater than 100 mg Cl/l of drinking water; one of these communities (i.e., Burlington) had one well with greater than 200 mg Cl/l. Consequently, the salting of icy and snow-covered roads was suspended in

Burlington. Furthermore, the Massachusetts Department of Public Health has warned that 62 communities in the state have levels of sodium in the drinking water exceeding the American Heart Association's (AHA) recommended standard of 20 mg Na/l (Massachusetts Department of Public Health, 1969–1970). By 1975, greater than 90 Massachusetts communities had drinking water sources exceeding the recommended AHA standard (Massachusetts Department of Public Health, 1975). Certainly, more research is needed to investigate the effects of elevated sodium levels in drinking water on the development of hypertension in "normal" and high-risk individuals.

Infants, as a potential high-risk group to the toxic effects of sodium from both food and drinking water, have recently received a great deal of attention (Smith, 1974; Philpott, 1975; Robertson, 1975). Hypernatraemia, especially hypernatraemia associated with dehydration, has been found to be a cause of permanent brain damage and of mortality in young bottle-fed infants (Davies, 1973; Smith, 1974). This incidence of morbidity and mortality is thought to be caused by the excessive sodium levels in food made up with unmodified powdered cows' milk. A comparison of human breast milk with cows' milk reveals that cows' milk has nearly 400 percent more sodium (580 mg/l versus 150 mg/l). The exposure of the infant to high levels of sodium together with the inability of the immature kidney of the neonate to concentrate salt in the urine may result in the condition of hypernatraemia (Robertson, 1975).

Robertson (1975) has also indicated that in artificially softened water as compared with drinking water sources with natural salinity due to NaCl a different problem may develop. In this instance, the base exchange mechanism results in each calcium ion being replaced by two sodium ions. The net result is a weak solution of sodium carbonate which may ultimately lead to a degree of alkalosis whose symptoms include depressed respiration and tetany. In support of the research of Robertson, Philpott (1975) reported a high incidence of neonatal tetany associated with low blood calcium in a hospital with a water softener. He noted further that infant formulas made up with tap water brought from his home contained half as much sodium as formulas prepared by the same personnel using hospital softened water.

Trace Metals

In addition to mineralization as a factor in the accentuation of cardiovascular disease, the effects of trace elements, besides cadmium, must also be considered. Reviews by Masironi (1969) and Schroeder (1974) have provided a necessary perspective for trace metal considerations in cardiovascular deaths.

Dietary chromium deficiency has been implicated as an important contributory factor in the development of atherosclerosis. Rat experiments have indicated that chromium deficiency is associated with a higher incidence of aortic plaques. Additionally, the presence of chromium in the diet has been reported to prevent the development of atheromatous lesions, to reduce the blood cholesterol level, and to increase the life span of the test animals (Schroeder and Balassa, 1965; Schroeder and Buckman, 1967; Marisoni, 1969). Schroeder and Balassa (1965) have also demonstrated that chromium acts to promote cholesterol metabolism and excretion. Studies with humans have indicated an inverse relationship of chromium tissue levels and cardiovascular disease. Thus, North American populations who are known to have a high incidence of CVD have extremely low levels of chromium in their heart tissue, and Africans and Orientals who have low incidences of CVD have high levels of chromium in their tissues (heart, kidney). According to Schroeder (1967, 1974), the low levels of chromium in the Western diet may be caused by the high consumption of refined sugars, since refined sugars contain almost no chromium, whereas raw or dark-brown sugars have appreciable amounts of it.

Manganese has also been suggested as an important factor in preventing the development of atherosclerosis, based on studies with rabbits and atherosclerotic patients (Kolesnikov, 1958). Also, vanadium has been found to diminish the levels of cholesterol in plasma and in the aortas of rabbits and humans (Mountain et al., 1956). The antiatherosclerotic effects of vanadium seem to be due to both reduction in cholesterol synthesis (Curran and Azarwoff, 1958) and to the acceleration of cholesterol metabolism. Finally, Schroeder (1966) reported a significant negative correlation between the vanadium content of municipal water supplies and atherosclerotic heart death rates.

A series of reports by Kanabrocki et al. (1964, 1965) and Miller et al. (1967) has indicated a positive relationship between serum and urine copper levels and the occurrence of myocardial infarction. Research by Harman (1963, 1968) not only substantiated the research of Kanabrocki et al. but also showed that dietary copper was able to induce atherosclerosis in animals. Furthermore, soft water, which has been associated with the development of cardiovascular disease, has also been shown to have significantly higher levels of Cu than hard water (Marisoni, 1969). Since soft water may corrode copper pipes, it may be expected that the replacement of galvanized pipes containing cadmium with copper piping may not offer the hoped for cardiovascular benefits as originally expected.

In addition to the above-discussed trace elements which may affect the incidence of cardiovascular disease, the literature also indicates that other trace elements found in drinking water (e.g., arsenic, cobalt, fluoride, iron, selenium, silicon, and zinc) may also affect cardiovascular function. Table 17 summarizes the relationship of trace elements to cardiovascular functions.

At this time, it seems apparent that a relationship between cardiovascular disease and the levels of various minerals and trace elements in drinking water exists. Certainly, a clear picture of what precise relationships exist still remains to be developed from both epidemiological and toxicological studies. It must also be realized that CVD results from many interacting factors, including age, genetic background, diet, exercise, smoking behavior and stress as well as minerals and trace elements in water. Some of these factors are not readily amenable to control such as age and genetic background.

Table 17. Relation of Trace Elements to Cardiovascular Function

Elements	Alleged Beneficial Effects	Alleged Harmful Effects
Mn	Protects against atherosclerosis	
Cr	Protects against atherosclerosis	
V	Protects against atherosclerosis	Orally, induces atherosclerosis
Co	Protects against atherosclerosis	Orally, induces atherosclerosis
Zn	Protects against hypertension	
F	Protects against calcification of the aorta	Produces focal myocardial necrosis (in large doses); damages the heart and produces blood pressure changes
Se	Protects against cardiac necrosis	
Si	Maintains elasticity of blood vessels	
Cu	Maintains elasticity of blood vessels	Enhances atherogenesis
Cd		Induces hypertension and atherosclerosis
As		Induces focal myocardial necrosis (in large doses)

Source. Masironi (1969).

However, most other factors, including the drinking water quality, are subject to human control. It is interesting to note that in 1963 an article in *Consumer Bulletin* suggested that a person with a family history of heart disease should consider moving to a hard-water area (see Figures 23 and 24). However, Winton and McCabe (1970) criticized this suggestion as premature, since they felt the evidence was not sufficient to determine the optimum range of water hardness.

KIDNEY DISEASE

In addition to the influence of cardiopulmonary disease predisposing affected individuals to the harmful effects of air pollutants, kidney failure (see "Cystinuria," "Cystinosis," and "Tyrosinemia") is also

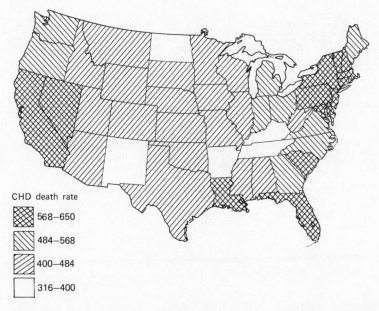

Figure 23. Average annual age = adjusted death rates for coronary heart disease (CHD), 1949 to 1951, by state—white men aged 45 to 64. *Source.* Winton, E. F., and McCabe, L. J. (1970). Studies relating to water mineralization and health. *J. AWWA*, **62**(1): 26–30. Reprinted from *Journal American Water Works Association* Volume 62 by permission of the Association. Copyright 1970 by the American Water Works Association, Inc., 6666 West Quincy Avenue, Denver, Colorado 80235.

known to exacerbate pulmonary infection. Animal studies have indicated that the bacterial clearance mechanisms of the lung are reduced following acute renal failure. Also, any kidney disorder would be expected to affect the "normal" urinary excretion of various pollutants. Such diseases may lead to the development of potentially dangerous accumulations of toxic substances in the body (Goldstein and Green, 1966).

Hanhijärvi (1974) reported that patients with severe renal insufficiency have significantly higher levels of free ionized plasma fluoride as compared with unaffected controls, when such individuals are exposed to levels of fluoridated drinking water containing either 1.0 or 0.2 ppm. Further studies revealed that the renal fluoride clearance activity is significantly lower in patients with renal insufficiency in both fluoridated and nonfluoridated areas. Similar results were found for individuals with nephritis (Hanhijärvi, 1974).

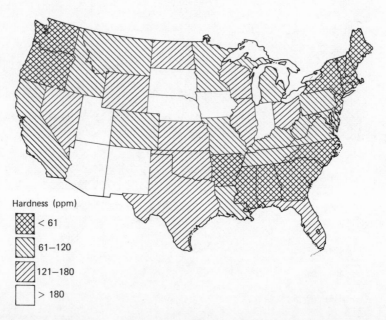

Hardness (ppm)

⊠ < 61

◩ 61–120

▨ 121–180

☐ > 180

Figure 24. Weighted average hardness, by states, of finished water from public supplies for 1,315 of the larger cities in 1952. *Source.* Winton, E. F., and McCabe, L. J. (1970). Studies relating to water mineralization and health. *J. AWWA*, 62(1): 26–30. Reprinted from *Journal American Water Works Association* Volume 62 by permission of the Association. Copyright 1970 by the American Water Works Association, Inc., 6666 West Quincy Avenue, Denver, Colorado 80235.

Individuals with diabetes mellitus and heart insufficiency who live in communities with fluoridated drinking water were also found to have their renal clearance of fluoride significantly impaired. However, the renal clearance of fluoride by individuals with diabetes mellitus was not significantly affected in a nonfluoridated community (Hanhijärvi, 1974).

In addition to affecting the levels of fluoride in blood plasma, the kidney has been found to play a critical role in the regulation of blood pressure levels, in part, by controlling sodium and water excretion (Freis, 1976). Freis (1976) presented evidence that in acculturated peoples the extracellular fluid (ECF) volume may be as much as 15 percent greater than that of unacculturated peoples ingesting much less salt and having significantly lower hypertension. Numerous other studies have indicated a relationship between ECF volume and the development of arterial blood pressure (Murphy, 1950; Watkin et al., 1950; Ledingham, 1953; Borst and Borst, 1963; Tobian, 1972). These animal and human studies have shown that a rise in extracellular volume is accompanied by increases in cardiac output and in blood pressure. The elevated extracellular volume then decreases while total peripheral resistance increases to maintain the hypertension. Therefore, the chronic state of hypertension is characterized by a normal cardiac output and an increased peripheral resistance. The precise manner by which the increased peripheral resistance is maintained is controversial and may involve both autoregulation (Fries, 1960) and humoral agents (Haddy and Overbeck, 1976).

An enhanced urinary output along with a decrease in ECF accompanies the increase in blood pressure. The common factor required for the development of any chronic elevation of blood pressure is for the kidney to increase urine volume and sodium excretion so as to prevent a chronically enlarged ECF (Guyton et al., 1974). The increased diuresis is directly affected by the capacity of the particular kidney to excrete an excess of sodium. The more efficient the functional capacity of the kidney, the more limited the rise in blood pressure, and the reverse is true (Guyton et al., 1974).

Other studies have revealed that different rat strains have been selectively inbred for both a predisposition and a resistance to hypertension. Transplantation of kidneys of hypertension-prone rats into the hypertension-resistant strain affects the development of hypertension in the host. The reverse experiment, with the hypertension-prone animal serving as the host, revealed that the presence of kidneys of a hypertension-resistant strain lowered hypertension (Haddy and Overbeck, 1976; Dahl, 1972; and Bianchi et al., 1974). These animal studies and the previously discussed human research indicate that kidney function, as affected by heredity and disease state, plays an important role in the development of hypertension. Consequently, individuals with an impaired kidney function are at increased risk with respect to the development of hypertension, especially if they ingest quantities of sodium greater than required for normal physiological function.

LIVER DISEASE

Liver damage which impedes the development of detoxification enzyme systems (see "Immature Enzyme Detoxification Systems") would also be expected to modify one's susceptibility to various pollutants. Activation of the liver microsomal detoxification system plays a significant role in the detoxification and excretion of a broad variety of environmental pollutants, including numerous drugs, alcohol, PCBs, and many insecticides. Liver microsomal enzymes may also convert inactive substances to their functional or active state. In a number of instances precarcinogens are converted to the carcinogenic state (Ariens et al., 1976) (see also Table 25, p. 180). Consequently, whether liver damage either enhances or diminishes pollutant toxicity is dependent on the specific pollutant.

Behavior and Pollutant Exposure

6

SMOKING

Pulmonary Effects and Heavy Metal Exposure

Certain behaviors tend to bring the individual into greater than "normal" exposure to specific pollutants. Often the behavior is intimately or inseparably related to the pollutant exposure. The case of smoking behavior is a prime example, since the act of inhaling tobacco smoke (part of the actual smoking behavior) brings about an additional exposure to various pollutants. As is commonly known, cigarette smoke has been experimentally shown to have a paralyzing effect on the cilia of the lung (Carnow, 1966). In addition, it may also cause a proliferation of mucous glands, leading to a marked thickening of the mucous blanket so that cilia drown in the mucous blanket. This tends to prolong the period of contact of the irritant with the bronchial wall. Also, acute inflammatory alterations, secondary to the adhering irritants, may occur, with breakdown of resistance of the bronchial wall and possible invasion of pathogenic bacteria. Thus it is quite clear that smoking behavior interferes with the normal cleansing mechanisms of the lung. It may also increase pulmonary resistance because of a narrowing of the bronchial passage. This narrowing is accentuated by the increased thickness of the mucous blanket, which is secondary to the bronchial irritation.

Cigarette smoke is also known to contribute considerably to the body's exposure to carcinogenic hydrocarbons and radioactive elements (see Tables 18 and 19) as well as numerous heavy metals, including As, Cd, F, Pb, Ni, Zn, and others (Menden et al., 1973). Analyses of cadmium levels in cigarettes revealed that 30 μg Cd is present per pack of which 70 percent passes into the smoke (Nandi et al., 1969). In fact, necropsy studies indicated that Cd levels in organ tissues of smokers are related to the number of packs of cigarettes smoked per year (Lewis et al., 1972). It is interesting to note that a proposed ambient standard of 0.05 μg Cd/m^3 would permit a total body accumulation of Cd of approximately 5.0 mg/45 years. In contrast, if a person smoked one pack of cigarettes each day for 20 years, the net contribution of Cd to the body would be approximately 5.0 mg, assuming 0.067 μg Cd/cigarette, or equal to that permitted by the proposed standard (EHRC-Cadmium, 1973) (Figure 25). The philosophical question of whether standards (air and water) should factor in the exposure of the smoker to cadmium is raised in the March 14, 1975, Federal Register with regard to the drinking water standard for cadmium.

Menden et al. (1973) reported that there is 4.25 to 7.55 μg Ni/cigarette and that although only 0.4 to 2.4 percent of the Ni in the smoked part of

Table 18. Active Components of Cigarette Smoke

Gases	Particles
NH$_3$	Acids; alkanes and alkynes
NO$_2$	Alcohols and esters; aldehydes, ketones, and quinones
HCN	Aliphatic hydrocarbons; aromatic hydrocarbons
HCHO (formaldehyde)	Sterols; nitriles, cyclic ethers, and sulfur compounds
CH$_2$ = CHCHO (acrolein)	Alkaloids; brown pigments; phenols (chlorogenic acid and other polyphenols)

Source. Kilburn (1973).

the cigarettes was present in the particulate phase of the mainstream smoke (i.e., the smoke inhaled directly by the smokers), between 11 and 33 percent of the Ni in the smoked part of the cigarette was present in the side-stream smoke. It was concluded that Ni in the side-stream smoke of cigarettes may be a health hazard to nonsmokers as well as smokers. Similar trends in percentages were found for Cd (1.56 to 1.96 μg Cd/cigarette; 7 to 10 percent Cd in mainstream smoke; 38 to 50 percent in sidestream smoke). However, as for Zn and Pb, considerably

Table 19. Identification of Suspected Tumorigenic Agents in Cigarette Smoke

Type of Components	Estimated Concentration in 100 Cigarettes (85 mm, Nonfilter)	Relative Importance in Experimental Tobacco Carcinogenesis
Carcinogens and tumor initiators:		
Polynuclear aromatic hydrocarbons[a]	10–30 μg	Major tumor initiators
N-Heterocyclic hydrocarbons	1–2 μg	Minor importance as initiators
N-Nitrosamines	10 μg	Suspected carcinogens of some importance
Nitroolefines	1 μg	Suspected carcinogens of minor importance
0, p'-DDD	10–100 μg	No essential contribution
Maleic hydrazide	10–100 μg	No real importance suspected
β-Naphthylamide	2–3 μg	Suspected bladder carcinogen
Other aromatic amines	10–50 μg	Carcinogens of some importance
Polonium-210	1–5 pc	Of some importance
Promoting agents:		
Neutral promoters (unknown structures)	?	Of essential importance
Volatile phenols	20–30 mg	Of some importance
Nonvolatile phenols	?	Possibly of some importance

[a] Four- and five-ring aromatic hydrocarbons.

Source. Wynder and Hoffmann (1968). Experimental tobacco carcinogenesis. *Science* **162**: 862–871. Copyright 1968 by the American Association for the Advancement of Science.

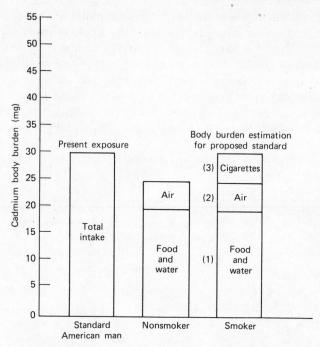

Figure 25. Contribution of various sources to cadmium body burden of man at age 45: (1) based on 1.25μ cadmium uptake for food and water per day; (2) based on 0.3 μg cadmium per day uptake from air (0.05 μg/m³); (3) based on 0.67 μ cadmium uptake from 20 cigarettes per day/20 years. *Source.* Environmental Health Resource Center, Illinois Institute for Environmental Quality (1973).

smaller percentages were found in the mainstream smoke, and levels in the side-stream smoke were even more reduced.

It is also important to note that levels of lead-210 and polonium-210 of smokers have been found to be about twice those in nonsmokers both in ribbones and in lung tissue (Holtzman and Ilcewicz, 1966). If the ribs were taken to represent the skeleton, Holtzman and Ilcewicz calculated that the radiation dose to the smoker from cigarette smoking would amount to 50 mrem/year of the average total skeletal dose rate of approximately 150 mrem/year, or approximately 33 percent of the total. They suggested that in the development of epidemiological studies of low-level radiation (e.g., [226]Ra in drinking water) exposure, consideration should be given to smoking habits, since the ^{210}Pb series contributes a significant percentage of the total skeletal dose.

Pulmonary Effects and Carbon Monoxide and Nitrogen Dioxide Exposure

Carbon Monoxide and Smoking

Carbon monoxide (CO) levels in the body as measured by carboxyhemoglobin (COHb) levels are effected by both endogenous and exogenous factors. For example, people are constantly exposed to endogenous CO, which is formed primarily from the normal catabolism of hemoglobin, with a small part contributed by the breakdown of nonhemoglobin heme (Sjöstrand, 1949; Coburn et al., 1963). This results in a COHb level of approximately 0.4 to 0.7 percent. In addition, it is known that hypermetabolism, certain types of drugs, and hemolytic anemia may significantly increase normal endogenous CO production. For example, Coburn (1970) has reported that phenobarbital and diphenylhydantoin induce hepatic heme, and, after "induction," catabolism of hepatic heme amounts to over 50 percent of the total endogenous CO production. Concerning patients with hemolytic anemia, COHb levels may increase to 4 to 6 percent.

As for exogenous CO exposure, the majority of human exposure results from the incomplete oxidation of carbon-containing material. The most heavily exposed people in a nonoccupational setting are tobacco smokers (Stewart et al., 1974; Cohen et al., 1971; PHS, 1968). Cigarette smoke as inhaled contains 200 to 800 ppm CO, and smoke from cigars or pipe tobacco contains much more (Patty, 1962). The typical one-pack-per-day cigarette smoker is usually found to have a COHb level of 5 to 6 percent, and those smoking two to three packs per day average 7 to 9 percent COHb saturation. Stewart (1976) has indicated that COHb saturation resulting from tobacco smoking is additive to that resulting from other exogenous CO sources encountered in the environment. Thus, a one-pack-per-day cigarette smoker in Milwaukee has an approximate COHb saturation of 5.5 percent when nonsmokers in the same area had a 1.2 percent saturation. Similar differences between smokers and nonsmokers in various cities are also known.

One of the most important nonindustrial sources of CO in the environment is from the exhaust of automobiles, which contributes about 60 percent of the total CO emissions per year (Stewart, 1976). As a result of a national survey conducted in 1969–1972, it was revealed that 45 percent of the nonsmoking blood donors in 18 sections of the country, each one including both metropolitan and rural areas, exhibited COHb saturations greater than 1.5 percent and thereby clearly implied that the federal ambient CO standard was being regularly violated.

Table 20 represents the human response to various concentrations of COHb. These are the expected signs and symptoms of toxication for CO exposure. As either expressed or implied in the table, infants and patients with cardiovascular disease, anemia, lung disease, and

Table 20. Human Response to Various Concentrations of Carboxyhemoglobin

Blood Saturation (percent COHb)	Response of Healthy Adult[a]	Response of Patient with Severe Heart Disease
0.3–0.7	Normal range due to endogenous CO production; no known detrimental effect	
1–5	Selective increase in blood flow to certain vital organs to compensate for reduction in oxygen carrying capacity of the blood	Patient with advanced cardiovascular disease may lack sufficient cardiac reserve to compensate
5–9	Visual light threshold increased	Less exertion required to induce chest pain in patients with angina pectoris
16–20	Headache; visual evoked response abnormal	May be lethal for patients with severely compromised cardiac function
20–30	Throbbing headache; nausea; manual dexterity abnormal	
30–40	Severe headache; nausea and vomiting; syncope	
50	Coma; convulsions	
67–70	Lethal if not treated	

[a] Exposure to CO in concentrations in excess of 50,000 ppm can result in a fatal cardiac arrhythmia and death before the carboxyhemoglobin saturation is significantly elevated.

Source. Stewart (1975). Reproduced, with permission, from "The Effect of Carbon Monoxide on Humans," *Annual Review of Pharmacology,* Volume 15. Copyright © 1975 by Annual Reviews Inc. All rights reserved.

increased metabolic rate are considered at high risk to CO exposure.

Stewart (1976) has explained quite clearly the effects of CO on cardiovascular function. He reasons that every molecule of CO which passes through the lung combines with hemoglobin and thereby diminishes the oxygen-carrying capacity of the blood and causes a finite stress on man. He concluded that there is no dose of CO that is without an effect on the body. Whether or not this effect results in any clinical response is, of course, dependent on the dose of CO and the state of health of the exposed individual. The body usually adapts to this hypoxic stress by increasing cardiac output or by increasing blood flow to a specific organ. If the capacity to adapt is exceeded, tissue hypoxia ensues.

Ayers et al. (1970) have reported the effects of CO on myocardial function in healthy adults, patients with coronary artery heart disease, and patients with noncoronary heart disease. They indicated that in patients with no indications of coronary heart disease who were exposed to rapidly increasing COHb saturation of 9 percent over 21 to 120 sec, there resulted an increased coronary blood flow, increased oxygen extraction ratio by the myocardium, and an insignificant decrease in coronary sinus oxygen tension. In comparison with the patients with a noncoronary history, those with coronary heart disease did not develop a significant increase in coronary blood flow. Further, although the oxygen extraction ratio by the myocardium was increased, the coronary sinus tension decreased significantly. Also, significant decreases in the lactate extraction ratio and the pyruvate extraction ratio were noted. The authors concluded that a potentially serious condition could result from the inhalation of CO by a patient with coronary heart disease incapable of responding to the anoxic stress by increasing coronary blood flow.

Subsequent research by Aronow et al. (1972), Aronow and Isbell (1973), and Anderson et al. (1973) has verified the Ayers et al. conclusion. For example, these investigators have shown that patients with advanced coronary artery disease and angina pectoris experience a significant decrease in their exercise tolerance after exposure to low concentrations of CO (i.e., enough to increase their COHb saturation to 5 percent).

Stewart (1976) suggested that there may be no level of CO exposure which does not exert a significant stress on patients with advanced cardiovascular disease. He further suggested that it may actually be impossible to establish an ambient air quality standard for CO which would offer complete protection to each person, as the law now reads. He suggested that it may be more reasonable to identify those individuals at significantly increased risk to CO exposure and provide them with acceptable artificial environments rather than trying to lower the national CO standard.

Nitrogen Dioxide and Smoking

Nitrogen dioxide is an important pulmonary irritant which is found in extremely high levels in various types of tobacco smoke. For example, cigarette, pipe tobacco, and cigar smoke have been reported to contain approximately 300, 950, and 1200 ppm NO_2, respectively (Stokinger, 1962). As a result of the relative insolubility of NO_2 in water, the NO_2 is able to pass through the relatively dry trachea and bronchi into the moisture-filled aveoli of the lungs. On reaching the aveoli, the NO_2 is converted to nitrous acid (HNO_2) and nitric acid (HNO_3). It is important to realize that both nitrous and nitric acids are highly irritating and damaging to lung tissue.

Considerable information is known about the toxicity of NO_2 in animals, but information is more limited concerning its effects on man. In 1966, Cooper and Tabershaw published a comprehensive literature review on NO_2 toxicology, and Morrow (1975) has reviewed the post-1965 NO_2 toxicological literature. Freeman et al. (1968) found that continuous inhalation by rats of 2 ppm NO_2 for 3 days resulted in the following biological modifications: (1) loss of the cilia lining the bronchi; (2) changes in the epithelial cells of the bronchi; and (3) formation of inclusion bodies in the smallest bronchi. At concentrations as low as 0.5 ppm for 4 hr, Stephens et al. (1971) noted distention of the aveoli, with a tendency to develop emphysema, in rats. Furthermore, Freeman et al. (1964) reported that when animals are exposed to 0.5 to 25 ppm NO_2 for 3 months or longer, lung changes similar to those of human emphysema develop. Finally, other rat studies have indicated that NO_2 exposure significantly increases the activity of $GSH-P_x$ GSH-reductase, and GSH dehydrogenase within lung tissue. It should be noted that O_3 affects $GSH-P_x$ in a similar fashion (Chow et al., 1974). Human studies of both an epidemiologic and toxicologic nature have tended to support the previous animal research. Epidemiologic studies in Chattanooga, Tennessee by Shy et al. (1970, 1970a) and Pearlman et al. (1971) indicated that the incidence of bronchitis among infants and school children significantly increased when exposed to elevated levels of NO_2 (up to 0.1 ppm for a mean 6-month value). Other human studies indicated that normal males (18 to 40 years of age) experienced airway resistance to NO_2 exposure at 5 ppm after 10 to 15 min exposure.

Functional changes were not noticed below 1.5 ppm NO_2 (Abe, 1967; von Nieding et al., 1973). In patients with chronic nonspecific lung disease, airway resistance increased significantly above 2 ppm after less than 5 min exposure (Smidt and von Nieding, 1974). It is also interesting to note that normal humans exposed to 1 ppm NO_2 for 24 hr/day for 180 days developed blood biochemical alterations, including increased cholesterol, lipid, and lipoprotein levels. Such modifications suggest a possible relationship to arteriosclerotic actions (Kosmider and Misiewicz, 1973).

Based on the presently available data, it is possible to identify several definite high-risk segments of the population as well as a few potential ones. Thus the very young and the aged as well as those with cardiopulmonary disease are clearly at increased risk to NO_2 exposure. Furthermore, individuals with red blood cell defects such as G-6-PD deficiency may have a difficult time adapting to NO_2 oxidative stress, since NO_2 seems to act in a similar fashion as O_3 on the cell membranes (Fletcher and Tappel, 1972; Chow et al., 1974).

Lung Cancer as Influenced by Asbestos Exposure and Smoking Behavior

The most important cancer associated with asbestos workers is cancer of the lung. In numerous groups of asbestos workers, approximately 20 percent of all deaths are caused by lung cancers. This has been reported for both asbestos product factory workers (Selikoff et al., 1964) and among users of these products (Selikoff et al., 1973). Factors affecting the individual's response to the carcinogenic activity of asbestos include the age of the worker, duration of exposure, and smoking behavior. In 1968, Selikoff et al. found that lung cancer was not significantly increased in incidence among asbestos workers with no history of cigarette smoking; yet when such history was present, the incidence of lung cancer increased considerably over that expected among other cigarette smokers, in the absence of asbestos exposure. They concluded that an asbestos worker who smoked cigarettes had 92 times the risk of dying of lung cancer, as compared with like individuals who did not smoke cigarettes.

Larger studies by Hammond and Selikoff (1973) confirmed this finding. They reported that nonsmoking asbestos workers had few lung cancers but those who smoked had significantly more lung cancers than would have been predicted had they not been asbestos workers. It was determined that cigarette-smoking asbestos workers have approximately 8 to 12 times more risk of developing lung cancer as compared with nonsmoking asbestos workers (Table 21).

Finally, the relationship of smoking to coal miner's pneumoconiosis should be mentioned. Considerable evidence now exists to suggest that among coal workers there is a significantly higher incidence of emphysema and heart disease in those that smoke as compared with nonsmokers (Naeye et al., 1971).

Smoking and Vitamin C

Considerable evidence exists which has indicated that cigarette smoking lowers normal levels of vitamin C in the body. For example, Strauss and Scheer (1939) reported that cigarette smoking caused a constant and marked reduction in the excretion of vitamin C. This implied, according to the authors, that vitamin C may be being destroyed in the body by the components of the smoke. Furthermore, McCormick (1952) demonstrated by laboratory and clinical tests that smoking one cigarette inactivated in the body approximately 25 mg of ascorbic acid. A specific component of cigarette smoke, nicotine, when added to human blood reduced its ascorbic acid content approximately 30 percent. Other studies have revealed that there are significant differences in the ascorbic acid levels present in the milk of smoking and nonsmoking mothers (Andrzejewski, 1966). Also, statistically significant lower levels (dose-related) of ascorbic acid have been reported in both leukocytes and blood plasma of cigarette smokers as compared with nonsmokers (Calder et al., 1963). Supplementing the research of Calder et al. (1963), Brook and Grimshaw (1968) found that blood plasma ascorbic acid levels decrease as a function of age and smoking behavior and that heavy smoking had the same effect on blood plasma ascorbic acid levels as increasing chronological age by 40 years.

In more quantitative *in vivo* studies with human subjects, Pelletier (1968) indicated that ascorbic acid levels of the blood plasma of smokers were about 40 percent that of nonsmokers. After providing the subjects with 2 g of ascorbic acid daily, the levels of ascorbic acid for both groups stabilized at about the same levels. However, the urinary excretion of ascorbic acid by the smokers was lower than the nonsmokers. According to the author, this implied a greater use of the ascorbic acid by the smoker. Several other studies have confirmed the finding that smoking diminishes the vitamin C levels in the body (Goyanna, 1955; Dietrich and Buchner, 1960; Durand et al., 1962; Rupniewska, 1965; Schlegel et al., 1969).

Table 21. Expected and Observed Deaths among 370 New York–New Jersey Asbestos Insulation Workers, Jan. 1, 1963 to Dec. 31, 1973, by Smoking Habits

						Cause of death		
		Person-Years	Lung Cancer			Pleural	Peritoneal	
	Number of Men	of Observation	Expected[a]	Observed	Ratio	Mesothelioma	Mesothelioma	Asbestosis
History of cigarette smoking	283	2195	4.07	45	11.06	7	14	19
Current smokers	181	1443	2.48	32	12.09	6	7	12
Ex-smokers	102	752	1.59	13	8.18	1	7	7
No history of cigarette smoking	87	708	1.58	2	1.27	0	7	6
Never smoked	48	409	0.84	0	—	0	5	3
Pipe or cigar only	39	299	0.74	2	2.70	0	2	3

[a] Expected deaths are based on age-specific white male death rate data for the U.S. National Office of Vital Statistics from 1963 to 1971, disregarding smoking habits. Rates were extrapolated for 1972 and 1973 from rates for 1967 to 1971.

Source. Selikoff and Hammond (1975).

145

Finally, using data concerning serum vitamin C levels of a large number of individuals (greater than 4000) from the Canadian Nutrition Survey (Nutrition Canada, 1973) as well as their smoking status and vitamin C intake on the day before the blood samples were taken, Pelletier (1975) has reported that cigarette smokers have lower serum vitamin C levels than nonsmokers in confirmation of previously published reports (Pelletier, 1968, 1970; Brook and Grimshaw, 1968; Lewin, 1974) which were based on a limited number of volunteers. He reported an average 25 percent reduction in serum vitamin C levels for smokers of less than 20 cigarettes and approximately 40 percent for smokers of 20 cigarettes or more. According to the author, it was quite likely that the development of lower serum vitamin C levels in cigarette smokers is primarily a result of less vitamin C effectively available from intakes similar to those of the nonsmokers. Pelletier (1975) concluded by suggesting that the vitamin C requirements and intakes of cigarette smokers should be increased to compensate for this decreased bioavailability.

Comparison of Smoking with Polluted Ambient Air in Respiratory Disease

Waldbott (1973) has observed that, in attempting to recognize the specific health effects of various environmental pollutants from data derived from epidemiologic studies, numerous investigators have often neglected to consider cigarette smoking as a variable. Also, he has indicated that other studies have only considered oxides of sulfur without regard for the other toxic air pollutants (CO, NO_x etc.) that may also contribute substantially to morbidity and mortality. He further states that, in long-term studies on mortality of large populations, there have often occurred peaks both in death rates and in air pollution levels that were not related to sulfur oxides. Furthermore, Carnow (1970), in epidemiologic studies performed in the Chicago area, reported a significantly higher rate of acute respiratory illness (e.g., cough and shortness of breath) in smokers as compared with nonsmokers at increasing levels of SO_2. Carnow concluded that smokers constitute a higher risk group than even the elderly and the very elderly without cough and phlegm who were nonsmokers or mild smokers living in a large polluted city.

Finklea et al. (1974) have indicated that the adverse health effects from smoking are a stronger determinant than ambient air pollution with regard to influencing the development of chronic respiratory

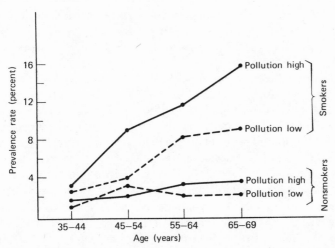

Figure 26. Age trends in respiratory disease in relation to smoking and air pollution. *Source.* Lambert (1970).

disease symptoms. In fact, the authors have estimated that the contribution of air pollution varied from one-third to one-seventh as important as that of cigarette smoking as a factor in chronic bronchitis prevalence. These estimates were based on both qualitative and quantitative differences in pollution profiles of different communities in addition to community differences in smoking patterns. Finklea et al. (1974) finally concluded that, although personal cigarette smoking is the most significant factor in the development of bronchitis prevalence, air pollution itself is a "significant and consistent contributing factor, increasing disease rates both in non-smokers as well as smokers from polluted communities." Figure 26, which represents an English study of 9975 men and women aged 35 to 69, clearly supports the conclusion of Finklea et al. (1974) and Carnow (1970) that smoking is a more important factor in the development of respiratory disease than typical urban air pollution levels, and yet indicates the definite contribution of air pollution to such disease incidence.

ALCOHOL CONSUMPTION

Alcohol in excessive amounts over a period of years is known to cause permanent damage to the liver. Any disruption of the liver has generally widespread health consequences, especially for the detoxification and excretion of foreign substances. Thus, it is expected that alcoholics may have their ability to detoxify foreign substances (e.g., DDT, PCB, etc.) impaired.

In a study designed to determine if consumption of alcohol predisposes one to lead poisoning, Cramer (1966) found a significantly higher incidence of lead poisoning among those who consume more than 75 cl alcohol per month. The author concluded that the alcohol consumption of lead workers should be determined prior to employment. According to Cramer (1966), those who acknowledge that they are regular alcohol consumers should be placed in the least dangerous position with respect to lead exposure (see "Lead Toxicity and Dietary Deficiency").

The role of alcohol consumption as a factor in the development of human cancers is a highly controversial area. Rothman (1975) has reviewed the literature with respect to the role of alcohol in carcinogenesis. Since alcoholic beverages often consist of a variety of biologically active substances, in addition to the ethanol, it is hard to define the effects of alcohol precisely. For example, many alcoholic beverages contain contaminants with a variety of aliphatic molecules with six or less carbons and one or more oxygen atoms, often in the hydroxyl form. Also, nitrosamines are known to be present in the maize beer of certain areas of Africa where esophageal cancer is frequent. Several researchers (Stenback, 1969; Kuratsune et al., 1971) have indicated that alcohol consumption increases the risk of cancer via its action as a cocarcinogen, that is, by enhancing or promoting the carcinogenic effect of other agents.

Rothman (1975) has noted that the effects of drinking alcohol are profoundly affected by smoking behavior. Table 22 illustrates that alcohol consumption not only confers an increased risk to oral cancer without accompanying smoking activity, but, in combination with smoking, a potent synergism results. The data indicate that heavy drinkers have a risk of 2 to 6 times greater than nondrinkers, depending on the degree of concomitant smoking activity. It should be pointed out that the 1.40 risk ratio for light drinkers who are nonsmokers may indicate an interaction of alcohol with nontobacco carcinogens or possibly a direct primary carcinogenic effect of alcoholic beverages.

Alcohol consumption has been associated with cancer at a variety of additional sites, including the larynx, esophagus, stomach, liver, and rectum (Rothman, 1975). However, in numerous instances, considerable controversy persists, and data from different studies often conflict with each other. At this time, further research is needed to more accurately define the role of alcohol in cancer development.

Table 22.   Risk Ratios for Oral Cancer According to Level of Exposure to Alcohol
and Smoking[a]

Alcohol per Day (oz)	Cigarette Equivalents per Day			
	0	Less than 20	20–39	40 or More
None	1.00	1.52	1.43	2.43
< 0.4	1.40	1.67	3.18	3.25
0.4–1.5	1.60	4.36	4.46	8.21
> 1.5	2.33	4.13	9.59	15.5

[a] Risks are expressed relative to a risk of 1.00 for persons who neither smoked nor drank.

Source.   Rothman and Keller (1972).

DRUG USAGE

Drug usage may also play a role, depending on the drug, in the response of an individual to various pollutants. Some drugs are known to stimulate the normal detoxification mechanisms, and other drugs are known to inhibit such processes. Some of the drugs affecting pollutant toxicity may be used only briefly in clinical treatment or over a prolonged time.

Of course, prolonged treatment with a substance which potentiates the toxicity of a pollutant would be quite dangerous. Because the possibility exists that commonly used drugs (depressants, narcotics, stimulants) may also interact with pollutants to exacerbate their toxic effects, there is a need for more research in this area (see "Immature Enzyme Detoxification Systems").

DIETARY PATTERNS RESULTING IN GREATER THAN AVERAGE POLLUTANT EXPOSURE

For a variety of reasons, certain pollutants are found in higher concentrations in some food than in others. This may be a result of natural biogeochemical cycles, as is the usual case with methylmercury in certain fish species, or agricultural practices and food processing activities together with the particular chemical properties of the pollutant and food type.

In addition to the high content of Hg in certain fish species, high levels of lead have appeared quite frequently in milk and fruit juice when contained in cans with a leaded seam (Mitchell and Aldous, 1974). The same authors speculated that if a child consumes canned products averaging 300 μg Pb/liter (50 percent above the mean level of their study), only 1 liter/day would be needed to meet the maximum daily permissible intake (MDPI) of lead set by an ad hoc committee of HEW and only 0.17 liter/day to exceed the WHO MDPI for a 10-kg child. Further studies based on assumed average daily diets for infants and the average level of lead in each food commodity group determined that children receive a higher exposure of lead per kilogram of body weight as compared with adults. This evidence coupled with the fact that infants and young children have higher absorption of Pb via the gastrointestinal tract than adults indicates the increased risk of children to environmental Pb poisoning (Kolbye et al., 1974).

Lipid-soluble pollutants such as DDT and PCBs are found in fatty tissue or substances with a high fat content. It has been reported that the PCB concentration in human breast milk from women in certain American cities is such that breast-fed children from these mothers would be exposed to more PCBs per kilogram of body weight than the average individual. PCBs are also known to concentrate in the fat tissue of various freshwater fish species such as salmon and trout (see p. 11 for a discussion of PCBs in infant diets).

People following certain ethnic diets or weight-watching programs which involve consuming large amounts of fish as compared with the general population would probably be exposed to higher levels of mercury and PCBs.

Screening Tests for Hypersusceptible Populations with Particular Reference to Worker Exposure

According to Mittman and Lieberman (1973), medical screening may be defined as the application of medical tests to large numbers of people to benefit those tested. Screening examinations generally produce new data that can improve the future health of the positive individual. The cost effectiveness of a screening test may be derived by calculating the cost of the test, the number of false positive and false negative results, and the benefits yielded by the use of the information in avoiding the precipitation of the diseased condition. The benefits may be viewed financially as it affects the individual, family, or society. Finally, realities such as pain and suffering must also be considered.

In addition to the identification of the specific genetic defect and the availability of a test which indicates the presence of the genetic defect, Stokinger and Scheel (1973) have indicated four additional factors that should be prerequisites for hypersusceptibility testing of workers. The genetic defect should (1) have a relatively high occurrence in the worker population; (2) concern pollutants usually encountered in industry; (3) be compatible with an apparently normal life until industrial exposure occurs; (4) be detected by an inexpensive test which can be relatively simply applied to the screening of large numbers of people.

Stokinger and Scheel (1973) have reported that five disorders fulfill the four prerequisites for the hypersusceptibility tests as just previously indicated. They include (1) SAT deficiency; (2) G-6-PD deficiency; (3) CS sensitivity; (4) reagenic antibodies to allergenic chemicals; (5) hemoglobin S in sickle-cell anemia. Other possibilities such as (1) atypical cholinesterases, (2) pyruvate kinase deficiency, (3) abnormal copper levels (Wilson's disease), and (4) abnormal porphyrin metabolism, according to Stokinger and Scheel (1973), failed to meet one or more of the prerequisites listed above. For example, detection of subnormal amounts of cholinesterase activity has not been associated with the health of pesticide workers, and the remaining three diseases are too rare to be considered.

The remainder of this chapter presents a review of the information concerning the development of medical screening tests for individuals thought to be genetically predisposed to toxic effects of various pollutants.

G-6-PD DEFICIENCY

For screening examinations, tests relying on visual indicators (i.e., dye decoloration, color development fluorescence) may be purchased from clinical laboratory suppliers. Beutler (1973) has described a simple fluorescence test which (1) bypasses the normal requirement of maintaining anaerobic conditions, (2) reduces time and manipulation, and (3) minimizes complications introduced when anemia is involved. For example, normal bloods show spots which fluoresce under long-wave light; the blood from G-6-PD deficients shows very weak, or the absence of, fluorescence. Table 23 provides an overview of the various tests for G-6-PD deficiency along with their advantages and disadvantages. Based on available evidence, it would seem that black males working in the welding industry (high ozone levels) and with hemolytic chemicals should be screened and monitored as a matter of policy.

SICKLE-CELL ANEMIA AND SICKLE-CELL TRAIT

Mass screening for sickle-cell anemia is actively carried out in the United States. This is particularly necessary in light of the vast numbers of individuals who may be carriers of the trait. There are 22 million black Americans in addition to people from Mediterranean and Eastern countries who are carriers. A screening test by Barnes et al. (1972) is known to be easily learned by inexperienced technicians; this test also distinguishes sickle-cell disease, sickle-cell trait, sickle-cell hemoglobin C disease, individuals with hemoglobin types AC, CC, and B_2, hereditary persistence of fetal hemoglobin F, and thalassemia minor at a cost of 5 cents per specimen. The timing of the initial screening is 10 min, but additional laboratory studies are needed to identify specific hemoglobin variants. Nalbandian (1972) has favored the use of the automated dithionite test as a preliminary screening method for the detection of HbS. As many as 960 blood specimens can be processed in an 8-hr technician workday at 2 cents per specimen by this method.

THALASSEMIA

Motulsky (1973) has indicated that most authorities recommend the use of electrophoretic techniques (various types of cellulose acetate elec-trophoresis) in β-thalassemia identification (see Barnes et al., 1972). With this methodology, some, but not all, cases of β-thalassemia trait will be determined (Motulsky, 1973). Thalassemia identification involves the counting of red blood cells and a hemoglobin determination along with electrophoresis with quantification for HbA_2 elevation. Serum iron determination, osmotic fragility, and red cell morphology tests are also needed. According to Motulsky (1973), there is no single test absolutely diagnostic for β-thalassemia, and the use of a variety of complementary tests is complicating, difficult, and more expensive. He also indicated a definite need for better techniques for determining thalassemia in order to make efficient application to mass screening.

CATALASE DEFICIENCY

Large-scale screening for the prevalence of catalase deficiency in man is well documented. For example, in Western countries the first

Table 23. Tests for G-6-PD Deficiency

Test	Advantages	Disadvantages	Reference
Heinz body test	Of historical interest only	Technically somewhat cumbersome	Beutler et al., 1955
Glutathione stability test	Reliable		Motulsky and Campbell-Kraut, 1961
Brilliant cresyl blue decolorization test	Relatively simple to perform	Anaerobic conditions required (capped with oil). Some lots of dye do not give reliable results. Prolonged incubation required; affected by anemia	
Methylene blue decolorization test	Relatively simple to perform	Must be gassed with CO; prolonged incubation required; affected by anemia	Toenz and Betke, 1962
DCIP decolorization test	Simple to perform, reliable, does not require anaerobic conditions	Prolonged incubation; affected by anemia	Bernstein, 1962; Scheuch and Kutscher, 1963
Methemoglobin reduction test	Inexpensive, reliable	Must be used on fresh blood or stored blood in special preservative; large volume of blood needed for most versions; spectrophotometer needed for most versions; prolonged incubation required	Dawson et al., 1958; Brewer et al., 1960; Bowman et al., 1964; Tizianello et al., 1966; Knutsen and Brewer, 1966

Test	Advantages	Disadvantages	References
Ascorbate-cyanide test	Relatively simple and inexpensive; quite sensitive for heterozygote detection	Large volume of blood required; relatively nonspecific; reading quite subjective and not always easy; prolonged incubation required	Rakitzis, 1964; Jacob and Jandl, 1966
Methylene blue absorption test	Inexpensive	Has not been extensively tested; more cumbersome than other modern procedures	Sass et al., 1966; Oski and Growney, 1965
MTT spot test	Relatively simple, reliable, very little blood required; rapid reading of results	Elution procedure time-consuming	Fairbanks and Beutler, 1962
Fluorescent spot test	Most specific and simple to perform; self-correcting for anemia over broad range, requires only 5 min incubation	UV hand lamp required	Beutler, 1971

Source. Beutler (1973). By permission of the author and the Israel Journal of Medical Sciences.

acatalasemic individuals were detected during the screening of 73,661 Swiss males reaching the age of 19 in 1961 and in 1965 (Aebi et al., 1961; Aebi, 1967). Furthermore, in order to determine the frequency of the acatalasemic gene in the Far East population, Takahara (1967), together with the Atomic Bomb Casualty Commission, investigated the frequency of hypocatalasemia in 82,969 Asian individuals, including Japanese and Koreans living in Japan, residents of the Ryukyo Islands, and Chinese residing in Taiwan.

For the determination of catalase in the blood, several rapid and accurate techniques are available (Aebi and Suter, 1969, 1972). In future screenings, the automated procedure as devised by Leighton et al. (1968) and Lamy et al. (1967) will probably be the chosen method.

PORPHYRIAS

Of the various types of porphyrin metabolic disorders the most frequent type is the hepatic porphyrias, that is, acute intermittent porphyria and variegate porphyria.

Acute Intermittent Porphyria

The levels of PBG in the normal individual's urine are too slight to be determined with the conventional method (Hammond and Welcker, 1948) described by Watson and Schwartz (1941). However, during an acute attack of porphyria, the levels of urinary PBG are quite high (Ackner et al., 1961; With, 1963). Additionally, the levels of ALA also increase significantly during an acute attack (With, 1963; Mauzerall and Granick, 1956; Ackner et al., 1961). Thus individuals experiencing acute intermittent porphyria can be detected. In latent porphyria, chromatographic analysis of the urine frequently (Stein et al., 1966) but not invariably (With, 1963) reveals increased excretion of ALA or PBG or both, thereby making confident diagnosis of asymptomatic individuals difficult (Watson, 1960; Waldenstrom and Haeger-Aronsen, 1963).

Variegate Porphyria

In patients with this disease, the concentrations of proto- and copro-porphyrin and peptide conjugates of dicarboxylic porphyrins are greatly increased in the feces even when clinical symptoms are minimal (Dean and Barnes, 1959; Eales, 1963; Sweeney, 1963). Although the frequency of variegate porphyria is not generally known, research does indicate an estimated 3 cases per 1000 white South Africans (Dean, 1963).

SCREENING FOR SUSCEPTIBILITY TO DRUG REACTIONS IN PATIENTS WITH CHOLINESTERASE VARIANTS

Study of the majority of patients with genetic abnormalities of serum pACHase has indicated the existence of a number of genes at the Ch_1 locus. These are referred to as the typical or normal gene (Ch_1^u) and variant or mutant genes which are characterized by a modified sensitivity of the pACHase enzyme to inhibitors such as dibucaine, fluoride, R02-0683, and butanol. The specific atypical genes are designated as the dibucaine-resistant (Ch_1^a), fluoride-resistant (Ch_1^f), and silent gene (Ch_1^s), which is associated with complete absence or a minute amount of activity in the homozygous condition (Lehmann and Liddell, 1964; Anonymous, 1968; Giblett, 1969).

Ten different genotypes of pACHase have been identified in man. Of these, four are known to be factors in affecting the sensitivity to suxamethonium, the muscle relaxant often used by anesthesiologists. These are the homozygotes for Ch_1^s and Ch_1^s and the heterozygotes $Ch_1^a Ch_1^s$ and $Ch_1^a Ch_1^f$. Their total or combined frequency in Europeans is calculated to be 1:1250. Intermediate sensitivity is known to occur in individuals with genotypes $Ch_1^f Ch_1^f$ and $Ch_1^f Ch_1^s$ with a calculated frequency of approximately 1:15,000 in Europeans (Anonymous, 1968, 1973). The usual techniques for phenotype (variant) identification use the benzoylcholine assay with dibucaine (Kalow and Genest, 1957) and fluoride (Harris and Whittaker, 1961). Also, the use of altered sensitivity of cholinesterase to inhibitors like R02-0683 (Liddell et al., 1963; Lehmann and Liddell, 1972) and butanol (Whittaker, 1968) also improves the identification of some genotypes.

According to Szeinberg (1973), several potential broad screening techniques have been developed which may be used for the detection of individuals homozygous or heterozygous for the variant genes, including those homozygous for the silent gene. These screening techniques include:

1. Agar diffusion and paper spot tests (Harris and Robson, 1963)
2. Test-tube method (Morrow and Motulsky, 1968)
3. Automated methods (Boutin and Brodeur, 1969)

Simpson and Kalow (1965), using the agar diffusion technique, developed a mass screening program which tested 6500 Brazilians. They reported only 0.1 percent false negative determinations but considerably more false positive results. Szeinberg et al. (1972) screened 9500 adult males with a slightly modified spot test and a follow-up confirmation by spectrophotometric procedures in cases where a positive determination occurred. They indicated that the spot test was easy to conduct and also inexpensive. However, the verification of positive determinations is more difficult and more costly.

Szeinberg (1973) stated that even though each case of prolonged apnea, caused by the presence of atypical pACHase in association with suxamethonium, causes a strain on the medical staff during operation procedures, no death connected with a single dose of suxamethonium has been reported. He further indicated that in well-equipped operating rooms little danger exists with regard to pACHase variant patients. Furthermore, Goedde et al. (1968) indicated that enzyme replacement has been successfully carried out by the injection of purified pACHase. Consequently, Szeinberg (1973) concluded that screening of every patient prior to suxamethonium does not appear justified.

SAT DEFICIENCY

Based on information previously discussed, it can be concluded that SAT deficiency satisfies the first three conditions for hypersusceptible screening. The requirement for test simplicity seemed satisfied by 1966 (James et al., 1966); however, tests of further sophistication were developed as a result of (1) the increasing number of mutants being reported and (2) the significant differences in the activities of body proteases that modified the expected phenotypic responses in SAT deficiencies (Stokinger and Scheel, 1973).

Initial tests measured the relative deficiencies of SAT (James et al., 1966; Smith, 1972; Briscoe et al., 1966). However, these tests did not indicate SAT variants. The new, more exacting tests detect the different genetic variants and now are a better predictor of the phenotypic response (chronic obstructive pulmonary disease—COPD) (Fagerhol and Braend, 1965; Fagerhol and Laurell, 1970). Lieberman and Mittman (1972) have developed a simple screening test that indicates abnormal SAT variants without regard for the level of antitrypsin protein in the serum. This is of importance, since various factors such as estrogenic agents and diseases may cause the levels to be elevated. Thus this test detects most of the SAT variants without regard to confounding variables.

Resnick et al. (1971) have reported the use of the SAT deficiency test in an industrial situation. They reported that of 89 coal miners with and without chronic respiratory disease just under 30 percent had below normal trypsin inhibitory capacity.

CYSTINURIA

Every patient with urinary calculi or with urinary tract sensitivity indicative of calculi should be examined for the possibility of cystinuria (Thier and Segal, 1972). The cyanide-nitropussitude test has been used as a chemical procedure (Brand et al., 1930). This test usually allows for easy detection of the homozygote; some, but not all, heterozygotes may be identified via this chemical test. More sophisticated and quantitative tests, if necessary, are available. (See Thier and Segal (1972) for a further discussion of identification tests.)

CYSTINOSIS

According to Schneider et al. (1967), fibroblasts derived from cystinotic patients and maintained for several generations have greatly increased quantities of free cystine. In fact, the levels of cystine from nephropathic cystinotic patients may have greater than 100 times the normal level of free cystine, and patients with "benign" cystinosis have approximately 50 times the normal free cystine content (Schneider et al., 1967, 1968). Schneider and Seegmiller (1972) suggest that these results imply that there is a quantitative and, therefore, an identifiable biochemical difference between the clinical expression of these two cystinotic variants, as well as the normal segment of the population. Although there is occasional overlap between the control and heterozygote for the nephropathic conditions, the differences between these groups remain highly significant for leukocyte ($P < 0.001$) and fibroblast ($P < 0.01$) values. The authors suggest that, since the cystine levels of heterozygote tissues are substantially less than half those present in tissues of homozygote patients, the cystine level is not directly proportional to gene dosage in this recessively inherited disease. Finally, individuals with benign cystinosis are usually identified only by "accident" during a normal ophthalmologic examination with a slit lamp due to cystine crystals in the cornea (Schneider and Seegmiller, 1972).

TYROSINEMIA

In the chronic stage, tyrosinemia is clearly diagnosed by liver cirrhosis, methioninemia, tyrosyluria with p-hydroxyphenyllactic acid as the major metabolite, generalized amino aciduria of a specific form, glucosuria, proteinuria, and hyperphosphaturia with rickets. The acute stage is harder to identify, since fructosemia appears similar to transitory hypertyrosinemia combined with liver disease. Thus in order to distinguish tyrosinemia definitely, it is necessary to perform fructose, galactose, ascorbic acid, and phenylalanine loading tests. No biochemical test has been developed to identify heterozygous carriers of the disease (La Du and Gjessing, 1972; Bodegard et al., 1969; Lindemann et al., 1970).

WILSON'S DISEASE

Attempts at identifying asymptomatic carriers of Wilson's disease have produced ambivalent results. For example, a decreased ceruloplasmin level, a common biochemical abnormality in people with Wilson's disease, has been present in some heterozygotes and not in others (Bearn, 1953, 1972; Sternlieb et al., 1961). A procedure for identifying heterozygotes without a liver biopsy has been suggested by Sternlieb et al. (1961). The proposed method relies on the ratio of Cu^{64} incorporation into ceruloplasmin at 48 hr to the level attained at 1 to 2 hr after oral ingestion of Cu^{64}. Ratios of Cu^{64} at 48 hr to Cu^{64} at 1 to 2 hr of less than 0.559 and greater than 1.253 indicate the heterozygote and the homozygote normal, respectively, with 99 percent confidence. However, the ratio in the control group seems to be correlated with age and makes the determination of the heterozygote less certain (Bearn, 1972).

HYPERBILIRUBINEMIA: FROM THE CRIGLER-NAJJAR SYNDROME TO GILBERT'S SYNDROME

Since cases of hyperbilirubinemia are often identified in families, it has been suggested that it is of genetic origin. The precise modes of inheritance for the various disorders, including the Crigler-Najjar syndrome, are uncertain. However, Arias et al. (1969) have suggested that, in cases of intermediate hyperbilirubinemia, the genetic transmission is according to an autosomal dominant gene with incomplete penetrance. Difficulty in determining the basis of inheritance is compounded by the problem of identifying carriers of the trait who do not exhibit hyperbilirubinemia. According to Arias (1962) and Arias et al. (1969), the oral menthol tolerance test, even though widely used, is not successful in differentiating between icteric patients and anicteric family members thought to be carrying the abnormal gene. Thus the lack of differentiability by the menthol tolerance test severely restricts the degree of its diagnostic usefulness. Dutton (1966) indicated that the discrepancies with the extent of glucuronide synthesis with menthol and with bilirubin may not be unusual at all in light of the suspected variety of enzymes that may catalyze glucuronide synthesis.

With regard to mild cases of hyperbilirubinemia, there may be several unrelated causes such as infectious hepatitis (Arias, 1962; Hult, 1950) in addition to any genetic basis (Gilbert's syndrome). Individuals with Gilbert's syndrome are usually identified by satisfying the following criteria (Powell et al., 1967): chronic, mild, unconjugated hyperbilirubinemia with normal values for direct-reacting bilirubin; normal erythropoiesis and red blood cell survival; and absence of histologic or functional abnormalities of the liver and biliary tree. Based on limited familial studies, the data suggest an autosomal dominant type of inheritance (Powell et al., 1967).

IMMUNOGLOBULIN A DEFICIENCY

A large number of different blood donors (64,588) were tested for the presence of Ig A deficiency during the years 1971 to 1974. Ouchterlony's double diffusion test was used in the screening of Ig A–deficient blood donors with confirmation by a modified racket immunoelectrophoresis technique II. Additionally, hospital patients (9920) were also examined for Ig A deficiency by the double diffusion technique. Other Ig A deficiency technique measures (inhibition of hemaglutination and double antibody solid-phase radioimmunoassay) which are more sensitive than double diffusion may be also used (Koistinen, 1975).

IMMUNOLOGIC HYPERSENSITIVITY TO ISOCYANATES

According to Stokinger and Scheel (1973), two industrial plants in the United States which produce isocyanates have instituted preemployment screening tests. This screening program has been very effective in lowering the number of workers who normally would have experienced unnecessary clinical symptoms. Stokinger and Scheel (1973) provide a summary of the screening methodology.

CYSTIC FIBROSIS

The increased concentrations of sodium and chloride in sweat is the most constant and best defined abnormality which has been found in CF patients. A wide variety of screening techniques has been developed and are evaluated by Lobeck (1972).

The detection of heterozygotes is difficult because of occasional overlapping values in controls with regard to sodium and chloride concentrations in sweat. However, the presence of the cilia dyskinetic factor in sera and the observation of metachromasia in cultured fibroblasts of heterozygotes serve as an aid in their identification (Lobeck, 1972).

HYPERSUSCEPTIBILITY TO CARBON DISULFIDE

A reliable screening test does exist and is described on page 81.

ARYL HYDROCARBON HYDROXYLASE

Aryl hydrocarbon hydroxylase has been usually measured in humans by using mitogen-activated peripheral blood lymphocytes. However, according to Kouri et al. (1976), considerable variation exists with regard to AHH levels, depending on which particular parameter is selected on which AHH activity is based. For example, if AHH activity is considered on a per 10^6 cells, per milligram DNA, per milligram protein, or per cultured flask basis, the inherent variation for the assay for one individual on separate days is greater than 20 percent.

As a result of such inconsistencies in AHH activity determination, Kouri et al. (1976) have suggested that the AHH activity determination be based on the assay of another microsomal enzyme which is not affected by exposure to polycyclic aromatic hydrocarbons and whose activity is directly associated with the level of mitogen activation. They further suggested that the enzyme system NADH-dependent cytochrome C reductase may be such an appropriate enzyme. They based their suggestion, in part, on the following observations. The enzyme activity (1) can be determined in mitogen-activated cultured human lymphocytes, (2) has an absolute requirement for NADH, (3) is proportional to the number of mitogen-activated cells assayed, (4) has a pH optimum of 7.7, (5) is not affected by mitochondrial inhibitors or antibody to NADPH–cytochrome C reductase, and (6) is similar for 10 different people.

At this time, further research is necessary to delineate a biochemical technique for AHH activity more precisely. With the high degree of variation in presently used techniques and the research of NADH-dependent cytochrome C reductase only in the preliminary stage, serious thoughts directed toward human screening for AHH activity cannot be entertained.

Based on the requirements for screening tests as listed above, the following conclusions can be drawn:

1. Medical screening tests should be encouraged for deficiencies of SAT and G-6-PD, sickle-cell trait, and hypersensitivity to organic industrial chemicals (isocyanates).
2. Catalase and Ig A deficiencies, thalassemia, sensitivity to CS_2, and cholinesterase variants should be further studied in order to establish a more accurate assessment of the health risks from these conditions. Toxicity associated with catalase deficiency should be examined for possible exacerbation as a result of associated dietary deficiencies (vitamins C and E and selenium).
3. Deficiencies of GSH, GSH-P_x, GSSG-R MetHb reductase, the porphyrias, pyruvate kinase, sulfite oxidase, Leber's optic atrophy, and xeroderma pigmentosum are considered too rare to justify screenings.
4. The Crigler-Najjar syndrome is usually fatal in children, thereby not affecting many adults.
5. Wilson's disease, tyrosinemia, cystic fibrosis, Gilbert's syndrome, cystinuria, cystinosis, and PKU carriers do not have sufficient screening techniques that enable the heterozygote to be reliably determined. Also, more research is needed to assess the potential health risks associated with these conditions. Although the homozygous segment of these diseases can be identified, it is most often quite rare or so serious as to shorten life significantly on their own account.

6. The association of AHH inducibility with bronchogenic carcinoma must be reproduced in independent laboratories. Also, reliable assays of AHH activity remain to be developed.

This study is not intended to determine the cost-benefit analysis of the screening procedure but only whether there is an accurate analytical method for identifying homozygotes and heterozygotes for the different genetic disorders.

Perhaps the most articulate and reasoned perspective concerning the efficacy of hypersusceptibility testing in industry has been expressed by Cooper (1973):

> What is the current state of tests of hypersusceptibility? There is insufficient epidemiologic evidence to support the use of any of them as a criterion for employability, without many qualifications. On the other hand there is ample scientific evidence to support wider testing. Premature assumptions as to the necessity for such tests, or over-optimistic claims for their benefits, can actually impede testing. On the basis of what we now know, no employer should be regarded as liable or derelict for not choosing to screen his employees. If he screens all employees, he would have to consider whether he would be regarded as liable to criticism for using a positive test to deny employment, or conversely, for jeopardizing the health of an individual permitted to work with a positive test. If it is clearly understood that the appropriate application of tests of hypersusceptibility is still moot, still on trial, then progress can be made in studying them. To clarify this point, it would be well if NIOSH developed positions, including consideration of their possible value when appropriate in criteria documents developed for control of occupational exposures.

High-Risk Groups
in Perspective

The identification and quantification of groups of individuals at high risk to various pollutants leaves one to conclude that probably all people at some time during their lifetime will be at increased risk to one or more commonly encountered pollutants. It has been found that certain stages in our life cycle which all individuals must experience (prenatal, infancy, childhood, adolescence, old age) have a lowered adaptive capacity with respect to the toxic effects of certain pollutants. For example, embryos, fetuses, and neonates have "immature" enzyme detoxification systems which are thought to permit the unusual accumulation of certain pollutants, including PCBs. The young also are developmentally deficient in various immunoglobulins, especially Ig A, which is known to play an important role in resisting respiratory tract infection. Also, from puberty onward, the functioning of the body's cell-mediated immunity progressively deteriorates so that by "old age" the incidence of cancer becomes especially frequent. The exposure of individuals with a degenerative cell-mediated immunological capacity to carcinogens tends to predispose carcinogenesis in such individuals.

In addition to identifying particular states in one's ontogeny or life cycle which are susceptible to the effects of specific pollutants, each individual has daily weak points or times of greatest sensitivity or hypersusceptibility to the toxic effects of pollutants. Thus circadian rhythms play a factor in the development of infectious disease, carcinogenesis, and overall health status. These developmental (ontogenic) and circadian weak points in our life cycle and daily lives, respectively, are unavoidable. So, it is correct to say that every individual is at some point at increased risk.

However, superimposed on these developmental and circadian periods of hypersusceptibility are numerous other conditions which do not appear to be "universal" in their occurrence. These factors include one's familial or genetic background, dietary habits, disease state, alcohol and drug consumption, smoking behavior, and other factors which tend to be individualized matters.

For example, an individual with a G-6-PD or SAT deficiency has certain predispositions to the toxic effects of environmental pollutants in addition to those which everyone has. Although everyone is at increased risk at certain stages of development, these individuals are at even higher risk with regard to certain pollutants. This is not to imply that traits such as G-6-PD or SAT deficiency are entirely maladaptive. In fact, the high incidence of G-6-PD deficiency is theorized to confer enhanced adaptive capacity to such individuals when they live in a malarial infested environment. However, in a nonmalarial infested environment, which also has substantial quantities of photochemical oxidants present, individuals with a G-6-PD deficiency would have to be considered potentially hypersusceptible. Thus the adaptive capacity of any organism must be viewed in light of its environmental surroundings. Of particular importance with regard to enzyme deficiencies such as G-6-PD is the fact that the deficiency is present from infancy to adulthood. Not all enzyme deficiencies act in this manner, but it is important to identify the number of "person-years" affected by the genetic condition. Consequently, a G-6-PD–deficient person is always at higher risk to oxidants, regardless of age.

If a G-6-PD–deficient individual were also deficient to a certain extent in vitamins C or E or selenium, there would be an enhanced susceptibility to the toxic effects of ozone. In identifying the high-risk populations with regard to ozone toxicity, it is necessary to identify not only the frequency of the G-6-PD deficiency and vitamins C and E and selenium deficiency in the population but also those individuals who have both the enzyme deficiency and the nutritional deficiency. This combination of high-risk potential in an individual may be expected to have an additive or possibly synergistic effect on the toxicity of ozone. For example, the frequency of G-6-PD deficiency in the population is approximately 11 percent of American black males. Seven percent of the United States population have also been found to be "physiologically deficient" in vitamin E. Since American blacks have been constantly identified as having a greater frequency of nutritionally deficient individuals as compared with the general population, it could be reasonably assumed that the frequency of blacks with a vitamin E deficiency is substantially greater than 7 percent. However, for the sake of argument, let us use 7 percent as the percentage of blacks deficient in vitamin E. If 11 percent of black males are G-6-PD–deficient and 7 percent of these individuals are deficient in Vitamin E, then (0.11 × 0.07) 0.77 percent of the black male population would have both the G-6-PD and vitamin E deficiency. In a city the size of Chicago, with approximately 750,000 black males, approximately 5775 would have the dual deficiency. If these 5775 individuals lived in an environment with very low breathable ozone concentrations, they would have only a slight risk. However, if the ozone levels were high, as in Chicago, such individuals would become a very high-risk group.

Another situation can be used to illustrate how several factors may superimpose themselves on a particular high-risk group, resulting in a magnified risk. For example, children less than 5 years of age are known to absorb Pb via the gastrointestinal tract more efficiently than any other age group. With all other factors considered equal, the young should ingest more Pb per unit of body weight. However, in addition to absorbing Pb more efficiently, young children consume diets with larger amounts of Pb than the normal adult diet. Also, nutritional surveys of children in low-income groups have indicated that a large percentage of such children received significantly less than the recommended dietary allowances of calcium and iron. Deficiency of these two minerals is known to potentiate the toxic effects of Pb drastically. Consequently, it can be seen that dietary factors superimposed on a developmental predisposition to absorb Pb may create a new and even greater risk group than by just age factors acting alone.

Despite the fact that considerable attention was spent in developing a dose-response relationship with regard to ozone toxicity on G-6-PD–deficient individuals, these are theoretical associations. As such, theoretical calculations, although offering a new conceptual model from which to devise experiments, are, as the word states, theoretical. It seems evident that such deficient people are at increased risk. But how much? This question is critical because the federal EPA attempts to establish ambient air standards for "safe" levels of pollutants; without a sufficient data base, EPA is left with the unenviable task of trying to protect health and not unduly affect the economy often without enough information to support the decision realistically.

As previously stated, many of the associations discussed in this document are theoretical statements concerning the toxic effects of pollutants on the health of human high-risk populations. Theoretical in this sense often means that much of the supporting evidence is based on animal experiments and then extrapolated to humans. Unfortunately, in terms of our data base, decisions must be extrapolated from animal studies, *in vitro* and *in vivo* human studies, or epidemiologic studies. All these research efforts have their limitations. For example, animal studies may not simulate human condition closely enough, *in vitro* studies remove the study from the body as a whole to an isolated and less realistic representation of the actual situation, *in vivo* studies with healthy subjects do not adequately represent the high-risk segments, and epidemiologic studies may have too many confounding variables, which mask the real effect of the pollutant. However, despite the inherent limitations of each methodology, similar results from these fundamentally different approaches provide a strong indication of reliability and internal consistency.

In the many cases of human predisposition to the toxic effects of pollutants presented here, animal, human *in vitro* and *in vivo*, and epidemiologic studies are represented. Since this document is concerned primarily with high-risk individuals in the population, rarely do any of the studies truly represent this group. The reason is quite simple. In addition to the research limitations listed above, the ethical question of exposing high-risk individuals to potentially harmful pollutants has led many to conclude that such studies should not be conducted. Now, the rights of the general public and especially prisoners are more strongly represented. As a result of such factors, data which would provide confidence in standard setting are difficult to obtain

Despite limitations in the development of precise dose-response relationships of pollutants to health effects, a need exists to represent "potential" health consequences in order to make decision makers aware of the possible ramifications of particular actions. The fact that most of the chemicals which have been found to produce cancer in man are also carcinogens in animals strongly implies potential importance of animal studies as providing predictions of human carcinogenesis. Thus there is no need to apologize because a human high-risk group has been determined by inference from animal studies. At the same time, the need for more data must always be recognized. Only the awareness of potential health risks can provide decision makers with the necessary prudence to view potential danger situations in very circumspect ways. In fact, the characterization of high-risk groups is actually a specialized type of environmental impact statement with the emphasis on the "receptor."

As indicated earlier, all people have weak points at certain stages of their lives as well as at certain times of the day. Most people also have other factors making them additionally susceptible to various pollutants. For example, although the G-6-PD deficiency is most prevalent among black males, cystic fibrosis and SAT deficiencies are most common among Caucasians, and many other deficiencies seem to be equally shared by all races and both sexes. If one includes the widespread occurrence of additive or synergistic factors which increase the toxicity of pollutants, it is easy to see that high-risk groups are highly diverse and differentially susceptible among themselves, affecting many different individuals.

The case for decision makers including the potential or theoretical high-risk group in any standard derivation process has certainly been emphasized here. However, it is important to realize that a vast literature exists concerning not only theoretical animal model studies but clearly recognized human responses to both ambient and industrial pollutant exposures. Certainly the classic historical examples of such exposures are derived from the well-known pollution episodes of London, Donora, and the Meuse Valley, among others. These studies and others clearly illustrate the inherent susceptibility of the aged as well as those with chronic heart-lung diseases such as bronchitis, emphysema, asthma, and immunological hypersensitivity. They are the first to require hospitalization and die when relatively high levels of pollutants are encountered. It must be emphasized that this knowledge is no longer theoretical, nor based "just" on animal studies, nor just a small human exposure experiment in a rigidly controlled setting, but a real-life tragic occurrence. It should also be emphasized that in these pollutant episodes, the levels of any one pollutant were not sufficient of themselves to have been responsible for the tremendous increase in morbidity and mortality associated with these incidents. Thus, the additive and synergistic interactions of different pollutants, as previously mentioned, must also be recognized by decision makers. Although not every person develops some type of heart-lung disease, Table 26 (p. 187) indicates that phenomenal numbers of the general public do become afflicted with these diseases, for example, 4 to 6 million asthmatics, 6 to 10 million with chronic respiratory disease, and 15 million with heart disease severe enough to limit activity. The seriousness of these diseases as predisposing factors to the toxic effects of respiratory pollutants such as

NO_2, CO, O_3, SO_2 and particulate sulfates along with the significant numbers of individuals so afflicted clearly indicates the need to establish pollutant priority schemes, as discussed in the next chapter.

The effort made to quantify the specific high-risk groups is of particular importance, since it provides in more concrete terms a population subgroup that can be potentially dealt with. One of the weaknesses in the derivation of standards is the fact that whenever a cost-benefit analysis is determined, it is much easier to quantify the actual cost to industry, and thus the consumer, in terms of the actual expenditures for control devices but difficult to determine for cost in terms of human health. In addition to affecting the quality of human health in terms of degrees of comfort, there is also a physician-hospital cost in certain cases. This effort at identification and quantification of the number at increased risk to pollutants offers a step in the direction of providing decision makers with a more accurate appraisal of the health consequences of their proposed regulations.

Practical Applications and Pollutant Priorities

The identification and quantification of individuals at high risk to environmental or occupational pollutants is of profound significance for the determination of public policy with respect to the well-being of our economy and our personal health. Specifically, it provides a reasonable characterization of the type and number of people who would first experience morbidity and possibly death at particular levels of pollutant exposure. Such a characterization provides practical applications to deal with the numerous occupational and environmental health problems effectively.

ECONOMIC HEALTH ASSESSMENT

Knowledge of high-risk groups can provide a theoretical basis for preparing an economic assessment for environmental health regulations which are passed at both the federal and state levels. In fact, the Illinois legislature has recently instituted a law requiring that economic impact statements be developed for all environmental regulations passed by the Illinois Pollution Control Board (IPCB). Thus economic impact statements must be prepared for the recently promulgated state ozone episode regulations (IPCB, 1976) which require rather severe economic restrictions such as closing all parking lots with greater than 200-car capacity, closing all schools and federal buildings, restricting incineration of trash, and closing O'Hare International Airport if the ozone levels became dangerously high. Obviously, these actions have profound economic implications. Their financial cost should be incorporated into a cost-benefit ratio. However, the health cost must also be quantified not only in terms of the estimated number of illnesses but also the medical-financial cost if the standards are not achieved. Ultimately, the decision makers are looking for a precise dose response relationship which will provide a human, industrial, and medical cost at each level of exposure.

An interesting attempt, using high-risk groups to quantify the economic health cost resulting from environmental pollutants emitted from automobiles with a catalytic converter, was presented by Jaksch (1975). He estimated the levels of pollutant (sulfuric acid mists) emission from the automobile, the level of human exposure, dose-response relationships, days with exacerbated symptoms, number of illnesses, and the number of individuals affected per disease category (high-risk groups). These data were converted into number of physician visits, number of hospital admissions, drug consumption, and restricted activity days for each year under study. The net health costs attributable to the tenth model year are presented in detail, and the net health cost for the second and fourth model years are summarized. The research results indicated that by the second model year the catalytic converter would result in net national health damages of approximately $20 million/year. By the tenth model year, the damages would rise to approximately $480 million/year. Such attempts to estimate economic health costs resulting from technological innovations clearly indicate the need to describe accurately not just what types of individuals may be at increased risk to the "new" pollutants but also the number of individuals affected and accurate dose-response relationships.

ENVIRONMENTAL PLANNING: HIGHWAY SITINGS

Another practical application is the recognition of high-risk groups in preparation of sites for various major highways. An article by Wadden et al. (1976) reported a planning methodology which associates pollutant emissions from transportation systems with potentially exposed groups of humans. The basis of the methodology involves identifying both the total population and high-risk segments of populations with respect to the emission source. They identified high-risk zones as those with both a high percentage of their population at increased risk to a specific pollutant and a high emission level of the pollutant. By determining the number of hypersusceptible individuals in each high-risk zone, the number of zones, the total emissions of specific pollutants, and the population-weighted mean emission levels for every pollutant for the entire area in consideration, the basis of rating indices for comparing a set of alternative transportation plans should be provided.

Wadden et al. (1976) applied this methodology to a comparison of alternative highway proposals for a segment of northeastern Illinois. They determined the location and estimated concentrations of CO and Pb as well as noise levels for each alternative routing. Next, individuals considered at increased risk to these specific pollutants were identified and quantified. Table 24 indicates the potential high-risk groups and their specific stressor agents. Following the completion of these methodological procedures, Wadden et al. were able to present decision makers in Illinois with a series of reasonable options based on health considerations of the proposed highway alternatives. Furthermore, by following such a planning methodology, it is possible to expose potential "hot spots" of health concern which then can be avoided in the development of the study. Also, it is significant to point out that if the analysis used only CO as the pollutant parameter, many of the potential high-risk zones would not have become known. It should be recognized

Table 24. Populations at Risk

Susceptible Population	Pollutant Exposure
Children 0–2 yr old	Pb, CO, and noise and Pb and CO
Day-care centers	Pb, CO, and noise and Pb and CO
Hospital beds	CO
Long-term care beds	CO
Population over 65 yr old	CO
Elementary school population	Pb, CO, and noise
People below poverty level	Pb and CO

Source. Wadden et al. (1976).

that the work of Wadden et al. (1976) was primarily methodological and did not intend to identify and quantify all high-risk groups with respect to automobile emissions.

In Massachusetts, especially the eastern part of the state, very high levels of sodium are present in drinking water (Huling and Hollocher, 1972). This has resulted primarily from the salting of roads in the winter months in order to remove ice and snow. The application of Wadden et al.'s methodology in snowy regions should lead planners to consider highway location in relation to drinking water source.

Consequently, the approval of any highway construction plans should clearly consider a detailed population characterization of the area affected (the number of infants, children, women of childbearing age, etc.) as well as numerous other factors, including the physical changes that a new highway may create. Although gathering this health-related information would be an additional cost to industry and the general public, it most assuredly will lead to a more healthy human environment and possibly to reduced medical expenses.

NUTRITIONAL SUPPLEMENTATION PROGRAMS

The health status of every individual is dependent on the relationship of one's adaptive capacity to the environmental stressors encountered. Certainly, it is generally accepted that if one has a low "resistance" (i.e., adaptive capacity), the chances of contracting an infection are increased as compared with the person with high resistance to the stressing agent. The health status of an individual can, therefore, be manipulated by either modifying the strength of the stressing agent (e.g., changing the levels of CO, Pb, SO_2, etc.) or by modifying one's health status. Most planning strategies have focused on reducing exposure to the stressing agent. The development of federal air quality and drinking water standards seems to indicate the direct application of this approach. However, the second approach, that of modifying the adaptive capacity of the individual, is almost completely overlooked. Neither alternative is mutually exclusive. In fact, the two approaches are entirely complementary and should be used in the attainment of excellent health status. To claim that programs should be developed to increase one's adaptive capacity in no way implies that pollutant levels now presently encountered are acceptable.

This study has presented numerous examples of how dietary deficiencies may exacerbate the toxic effects of occupational and environmental pollutants. For example, relative deficiencies of Ca and Fe which are quite common are known to potentiate the toxicity of lead. Vitamin E has been found to reduce the toxicity of oxidizing agents (e.g., ozone) at levels presently encountered in the breathable atmosphere. Vitamin C has been found to reduce the toxicity of a variety of pollutants, including certain inorganic metals and organic compounds. It is quite apparent that the adaptive capacity of many individuals would be significantly improved by the consumption of nutritionally balanced diets.

It would seem that government-subsidized school nutrition programs could play a very important role in increasing the resistance of school-age children to environmental stressors. These programs are certainly important in assisting the normal physical development of the participating children. However, with the knowledge of dietary factors which influence pollutant toxicity, the importance of these programs is increasingly evident. The hot-lunch program for the elderly in certain parts of the country is another example of a governmental nutrition program which may have similar multiple health benefits. These programs are especially needed in the inner cities, where diets are notoriously poor and environmental pollutant levels (e.g., ozone) are usually high. Thus, in this instance, we can consciously affect the improvement of the adaptive capacity and thereby improve health status. Preventive medicine (e.g., proper nutrition in this case) will not only improve general well-being but also reduce health costs by diminishing respiratory infections, cancer development, and other deleterious responses. It is also important to realize that vitamins A and E are fat-soluble and that excessive supplementation is potentially dangerous, since, unlike water-soluble vitamins like the B complex and vitamin C, fat-soluble vitamins may be retained by the body and may reach potentially toxic levels themselves.

The entire area of diet-pollutant interaction is an important practical area which needs to be further explored. It is quite possible that dietary recommended allowances for certain vitamins, expecially the water-soluble vitamins, may be increased in order to reduce toxicity of environmental pollutants effectively. The next decade will hopefully yield some important research contributions to this area.

BIOMEDICAL IMPLICATIONS

From a biomedical perspective, the many theoretical high-risk groups are difficult to study because of the potential danger that such studies may cause the prospective subjects. Also, epidemiologic data, although very important in completing the total picture of how a pollutant may act, most often need to be supplemented by actual controlled laboratory studies. By coupling the new information about high-risk individuals with the need for controlled laboratory studies, it may be possible to stimulate the rapid development of animal models which simulate human high-risk conditions. Already several potentially workable models exist and may be used to provide more reliable extrapolations from animal studies to man. For example, the use of mice with low levels of G-6-PD (Hutton, 1971) and catalase deficiency (Feinstein et al., 1968) has already been reported. Gunn rats, which lack the enzyme glucuronyl transferase, may be a model which would characterize how individuals with the Crigler-Najjar syndrome and Gilbert's syndrome may respond to certain pollutants (Swarm, 1971). These models should be investigated to evaluate how well they simulate the human. Experiments with appropriate animal models, therefore, are vital to the continued development of more precise assessments of pollutant effects on hypersusceptible humans.

ROLE IN STANDARD SETTING

The application of the knowledge of high-risk groups should also play an important role in assessing the strategies which the federal and state EPA's use to reduce air and water pollutants. Presently, the federal EPA has established ambient air quality standards for six pollutants (CO, SO_2, NO_2, oxidants, hydrocarbons, and particulates) because of their toxic effects to humans and their ubiquity. For three other toxic but not as ubiquitous substances, Hg, Br, and asbestos, emission standards are applied.

The Clean Air Act specifically demands that primary air quality standards completely protect the public's health and that the standards have sufficient safety margins incorporated into them. According to Finklea et al. (1974), the Clean Air Act actually assumes that there exists a "no-effects" level for every pollutant and for each adverse health effect. However, individuals do exist who are more susceptible to the effects of environmental pollutants because of a variety of inherent biological factors acting either singly or in combination, and these individuals, termed high-risk segments of the population, are the first to experience morbidity or mortality from various levels of environmental pollutants. Figure 27 represents the spectrum of biological response within the population.

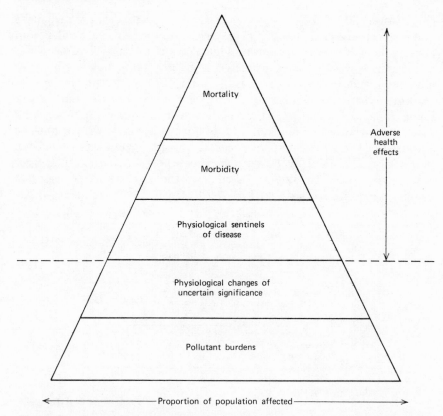

Figure 27. Spectrum of biological response to pollutant exposure. *Source.* Colucci, A. V., et al. (1973). Pollutant burdens and biological response. *Archives of Environmental Health*, **27**: 151. Copyright 1972–73, American Medical Association.

Threshold-Nonthreshold Controversy

Considerable controversy surrounds the concept of nonthreshold for a biological response to pollutant exposure. If thresholds were able to be determined, the establishment of environmental and occupational health standards would become a much easier task. It would certainly be quite reassuring to believe that all permissible levels of exposure are below levels which induce an adverse response during the individual's natural lifetime.

When discussing the threshold response concept, it is necessary to distinguish between the biological response on a population and on the individual. There is much less controversy when the term "nonthreshold" is applied to large heterogeneous groups of individuals. In this instance, it is easily recognized that there are spectra of responses and characteristics for nearly every trait from height, skin color, and general health status to immunological function. In a diverse population, there would not be expected to be any point below which an adverse response may not possibly occur. Thus, a relatively safe working hypothesis is that there is no threshold for pollutant effects on a heterogeneous population. There is, however, considerably less agreement over whether or not individuals have thresholds which must be exceeded before a toxic response occurs. This particular point has been previously addressed by numerous researchers, including Stokinger (1972). Stokinger considers toxicity as the net result of the two competing responses seen in Figure 28. On the left are depicted the

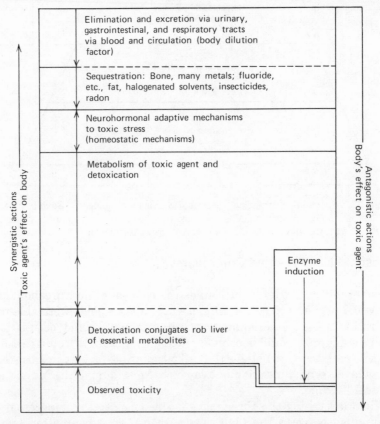

Figure 28. Analysis of the effects of a toxic substance. *Source.* Stokinger, H. E. (1972). Concepts of thresholds in standard setting. *Archives of Environmental Health*, **25**: 155. Copyright 1972–73, American Medical Association.

effects of the toxic substance on the body, and on the right are represented the body's adaptive or homeostatic processes which neutralize the pollutant's effect and return the system to equilibrium. Certainly the induction of liver microsomal enzymes, which help to detoxify foreign substances, and the production of antibodies represent examples of natural homeostatic, adaptive processes. Stokinger believes that a certain level or concentration of pollutant must be reached before the body's homeostatic responses can no longer overcome or neutralize the stressing agent. At this point, an individual threshold is said to exist. Thus, in the individual, thresholds for specific pollutants may exist. However, it should be emphasized that to recognize thresholds in individuals does not contradict the concept of nonthreshold responses when a large, heterogeneous group is considered.

Sometimes the liver "detoxification" enzymes "bioactivate" the substance in question to a more active state (Ariens et al., 1976). This is particularly true of some biochemical modifications by oxidation. Table 25 indicates six instances in which the microsomal enzymes induced bioactivation of the initial substance. As indicated in the table, several of these substances became active carcinogens. In these instances, the normal homeostatic responses are actually part of the toxification process. The concept of bioactivation by liver microsomal enzymes

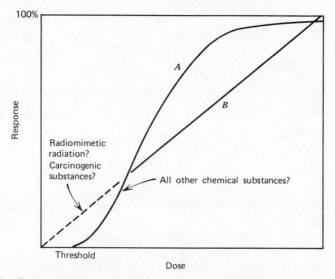

Figure 29. Dose-response relationship for a noncarcinogenic agent (A) and a carcinogenic (B) toxic substance. *Source.* Stokinger, H. E. (1972). Concepts of thresholds in standard setting. *Archives of Environmental Health*, **25**: 155. Copyright 1972–73, American Medical Association.

Table 25. Bioactivation and Biotoxification

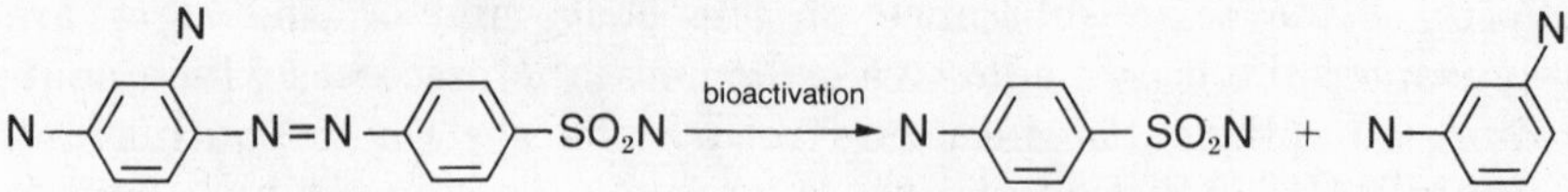

Prontosil rubrum
antibacterial *in vitro*

Sulfanilamide
antibacterial *in vitro* and *in vivo*

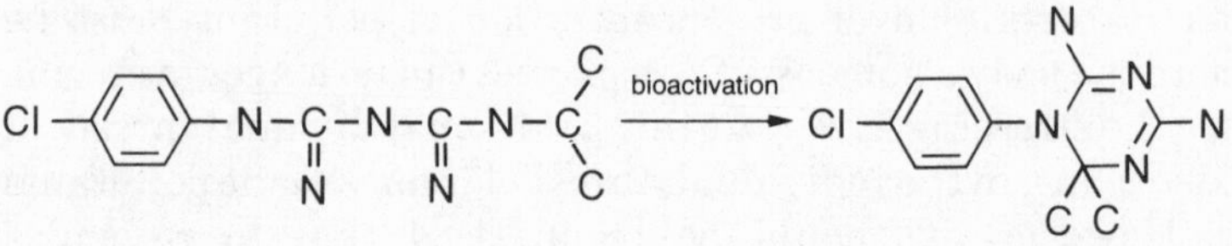

Proguanil

Cycloguanil
antimalarial

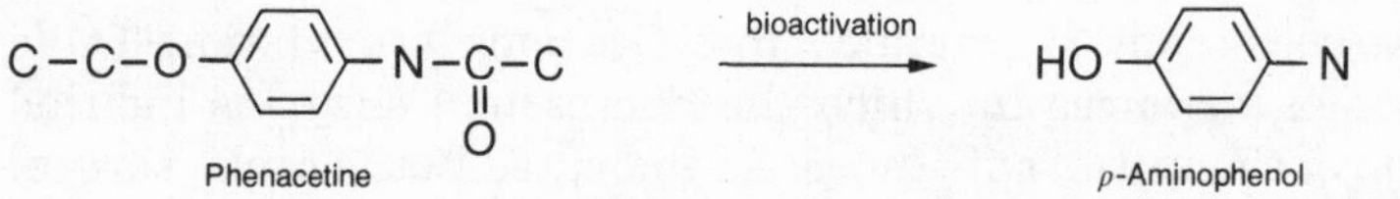

Phenacetine

p-Aminophenol
hemolytic

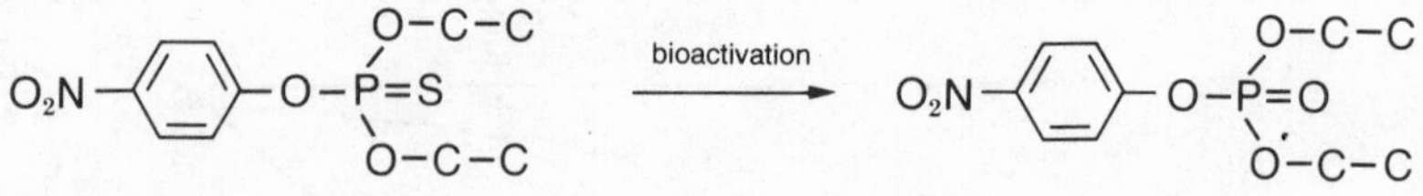

Parathione
weak ACh-esterase inhibitor

Paraoxon
strong ACh-esterase inhibitor

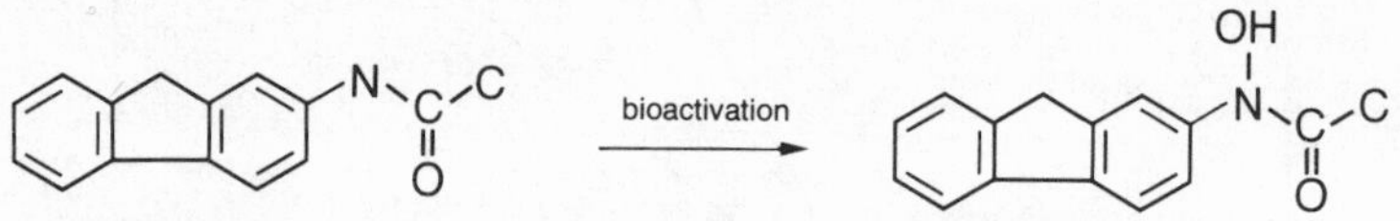

2-Acetyl-
aminofluorene
precarcinogen

N-Hydroxyacetyl-
aminofluorene
carcinogen

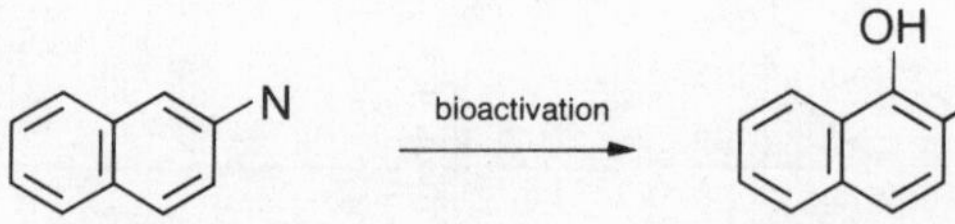

β-Naphthylamine
precarcinogen

α-Hydroxy-*β*-
naphthylamine
carcinogen

Source. Ariens et al. (1976).

raises serious questions to the validity of Stokinger's threshold response theory.

At this point, it is important to recall that there is considerable doubt whether individual thresholds actually exist for radiation damage and carcinogenesis (Figure 29). Ashford (1976) has indicated that, in contrast to the toxic effects of many chemical pollutants, the effects of radiation are cumulative and irreversible. He further states that each exposure to radiation causes some type of permanent damage which should be considered additive to the damage effected by previous exposures. From such reasoning, Ashford concludes that there is no threshold of activity for effects of radiation exposure. However, repair mechanisms do exist which may reverse some of the disruptive effects of mutagens such as radiation (Epstein et al., 1971).

Xeroderma pigmentosum, an inherited disease in humans thought to involve enzyme-deficient activity of an ultraviolet specific endonuclease, is characterized by an extreme sensitization of the skin to sunlight. Individuals with xeroderma pigmentosum lack or partially lack the ability to repair ultraviolet damage from sunlight, thereby associating a deficient DNA repair mechanism with enhanced susceptibility to ultraviolet toxicity (Cleaver, 1968). Such a deficiency clearly illustrates the lifesaving benefits of DNA repair mechanisms which may prevent a toxic response of ultraviolet at low exposure levels. Thus, in normal individuals without the ultraviolet specific endonuclease deficiency, every exposure to radiation would probably not result in permanent damage, in contrast to the previous suggestion of Ashford.

Jones and Grendon (1975) have derived a dose-response relationship for a variety of chemical carcinogens (hydrocarbons) as well as ionizing radiation. They indicated that the latency period shortens as the dosage increases according to an inverse-cube relationship. For example, if the dosage is decreased by a factor of 1000, there is a tenfold increase in the latency period and vice versa. Jones and Grendon suggest that if one extrapolates the time of appearance for cancers as a result of exposure from low doses of ionizing radiation, including radiation from ^{226}Ra, the induction time begins to exceed life expectancy. This would support Evans' (1967) "practical" threshold concept. Jones and Grendon based their prediction for the radiation-induced cancer relationship on animal studies (mice, ^{226}Ra, rats, X-rays; dogs, ^{226}Ra, ^{239}Pu, and ^{228}Th) and exposure of humans to ^{226}Ra. They concluded that the latent period appears to vary inversely with the 0.453 power of the dose which is fairly similar to the inverse–cube-root relationship noted above. Using this inverse relationship for a ^{226}Ra body burden of 0.1 pCi, the latent period for cancer induction would be 83 years. The examples of xeroderma pigmentosum and the research of Jones and Grendon illustrate the difficulty in attempting to define precisely the risks to human health of exposure to low levels of carcinogens.

A pollutant not presently considered a carcinogen, CO, has recently been suggested by Stewart (1976) as exhibiting a nonthreshold effect. Stewart has reported that every molecule of CO which enters the body displaces a molecule of O_2, thereby diminishing the O_2-carrying capacity of the blood. The body, however, adapts to the reduction in O_2-carrying capacity by increasing cardiac output or alteration in blood flow to various target organs. Such adaptational responses have been measured as low as 2 to 3 percent COHb in the blood. In the normal individual, the concept of homeostatic tolerance, as defined by Stokinger, is clearly in evidence. However, whether the response is adaptive or detrimental becomes dependent on the health status of the exposed individual. Previously discussed research (smoking and CO) has indicated that, in patients with coronary heart disease, there is a significant diminution of this normal adaptational response. In fact, Radford (1976) has reported that detectable effects can be measured on susceptible individuals as low as 2.8 percent COHb. According to Stewart (1976), this apparent 2.8 percent COHb "threshold" may be entirely due to the fact that the methodology to monitor effects below such a level does not yet exist.

Recent discussion has focused on several of the occupationally related implications of CO exposure (Stewart, 1976; see discussion section). For example, if recovered coronary patients return to work and are exposed to levels of CO between 25 to 50 ppm, what are their chances of experiencing another coronary attack and how could the role of CO in the development of the attack be interpreted? Also, what would be the legal implications? Even to attempt to answer such questions, it is important to recognize that, according to leading authorities, a person with far advanced coronary disease should not have a COHb level of greater than 5 percent. Yet, as previously pointed out, a sedentary smoker achieves such levels. Thus a person should be encouraged not to smoke if he was to return to a high CO environment, since both types of exposure are additive. However, as unpredictable as it seems, the epidemiologic evidence to date reveals that, for individuals who have previously experienced a heart attack, there is no difference in their chances of having a second heart attack whether or not they stop smoking. More information is needed to clarify this point, since it does not appear consistent with our knowledge of CO effects.

It is important to realize that when health implications of CO are considered, the major high-risk group which immediately stands out is not really those who have already had heart attacks but males over 45 years of age who have preexisting coronary disease that may be severe and who are unaware of it. This may include upwards of 25 percent of the males in this age group (Stewart, 1976).

A strong emphasis on some of the health effects associated with smoking has been presented here. The studies of Lambert (1970) and Carnow (1970) have clearly labeled smokers in a high-risk category toward the development of respiratory disease. Finklea et al. (1974) have indicated that smoking may be three to seven times more important than ambient air pollution in the development of respiratory disease. The important contribution of smoking to radiation and heavy metal exposure must also be recognized (Holtzman and Ilcewicz, 1966, and Menden, 1973). As a result of the acknowledged adverse effects of smoking on human health, it has been recently suggested that our national air standards for CO may really be misdirected, since we are avoiding the principal problem, the cigarette smoker (Stewart, 1976; discussion). Of course, this is a highly debated point, especially when the economic cost of implementation strategies is considered. Yet, the point remains that CO exposures are additive and that environmental exposure to CO in nonsmokers can result in COHb saturation levels of over 2 percent in the blood. Thus to diminish our concern for environmental quality because cigarette smoking may be a more significant factor with regard to some pollutants clearly misses the point and also clearly disregards human health considerations. Rather than abandoning our national ambient air quality standards, better educational programs are needed on the effects of smoking as well as governmental pressure on the tobacco industry to develop cigarettes which release reduced levels of carcinogens, CO, NO_2, NH_3, HCN, and metals during combustion.

Experiments by Buckley et al. (1975) concerning the toxic effects of ozone on humans also clearly indicate the efficiency of the homeostatic compensatory responses of normal individuals during ozone stress. Such people respond to the ozone (0.5 ppm for $2\frac{3}{4}$ hr) by increasing the activity of G-6-PD by 20 percent. This permits the normal individual to maintain sufficient GSH levels and thereby to ensure the integrity of red blood cell membranes. In contrast to normal individuals, people deficient in G-6-PD have significantly reduced abilities to generate sufficient adaptive responses. This explains why G-6-PD–deficient people are at such high risk to oxidant chemicals with regard to the development of hemolytic anemia. Such information clearly demonstrates that there are variable thresholds for pollutant effects in a population with a diverse gene pool such as ours.

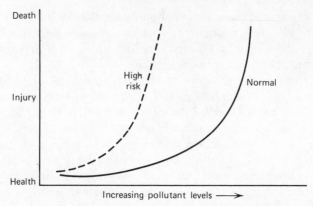

Figure 30. Comparison of the response of high-risk and normal individuals to increasing pollutant levels.

Figure 30 illustrates a theoretical comparison between normal and high-risk segments of the population with regard to the onset of toxic effects at increasing levels of pollutant exposure. Any normal adjustment or homeostatic adaptive response as well as compensatory capabilities are reduced in the high-risk group as compared with the normals. This results in the earlier onset of disease, disability, and death, depending on the specific situation. Also, the interaction of certain pollutants such as hemolytic oxidant-stress-type chemicals with G-6-PD–deficient individuals often leads to hemolytic anemia of varying severity, but only very rarely to death, since hemolytic crises in G-6-PD–deficient people are self-limiting diseases (see Beutler, 1972). In contrast, other pollutant-organism interactions may not be self-limiting, as those resulting in the cases of MetHb and COHb formation.

Safety Factors

So-called "margins of safety" should probably be considered as indicators of ignorance or, more generously, imprecise safety factors. Despite their enhanced susceptibility to the toxic effects of pollutants, high-risk individuals, according to Finklea et al. (1974), have not been specifically and separately considered in setting environmental health standards. The rationale is that adequate protection for the larger general population segments also ensures protection for the "large number of relatively small susceptible segments of the population for which we have little or no quantitative exposure information" (Finklea et al., 1974).

Thus, high-risk segments of the population were not specifically considered in the standard-setting process because there was not sufficient information about them to provide any quantitative assessment of risk and because it was assumed that they composed only a small percentage of the population. Despite the fact that safety factors recognize that certain people are more sensitive to pollutants than others, they are inherently imprecise. Often such safety margins include factors of 10 and 100 for variation within a species and between species, respectively. Undoubtedly, the precise difference in sensitivity between a statistically "normal" individual and one in the high-risk subpopulation is different for the various causes of the high-risk condition. In reality, it is an illusion to assume that a threshold exists in our diverse highly heterogeneous human population, although such an assumption may be of practical significance in many cost-benefit analyses. It cannot even be said that there are separate thresholds for the normal as well as the high-risk segments. Even among such a high-risk group as those with G-6-PD deficiency, there are now 80 recognized genetic variants (Beutler, 1972). Ultimately, each individual has his own unique genetic composition and, consequently, his own individual threshold.

Blanket safety factors of any designated number (e.g., 5, 10, 100, etc.) are usually intended to include all humans at increased risk to the pollutant exposure, but from the previous discussion it is apparent that such safety factors lack the precision of a quantitative analysis concerning the identification-quantification components and dose-response relationships. This is not intended to be critical of "safety factors" or the individuals who apply them. In fact, with the limited data that regulatory agencies often have, "imprecise safety factors" are the only realistic options short of completely eliminating human exposure entirely. This is intended, however, to point out deficiencies in our present system in order to improve the quality of our future health standards so that they more precisely define the actual protection they offer the whole population, including those at increased risk.

How High-Risk Groups Get Overlooked

Besides the inadequacy of applying imprecise margins of safety to protect the high-risk individuals, the standard-setting process has acted on an important false assumption. This assumption states that the high-risk segments of the population are only very small percentages of the total population. However, as discussed in Chapter 8, all people are at high risk at various points of their life cycle and even at specific times of their normal circadian rhythms. Table 26 indicates the estimated numbers of individuals at high risk to various pollutants from a variety of causes. The numbers are quite large, often encompassing significant percentages of the population of specific racial ancestries. To ignore hypersusceptible segments of the population in the development of health standards is actually to ignore everyone at some time during his ontogeny. Also, decision makers should not base health standards primarily on the results of experiments involving normal, healthy, active 25- to 30-year-old individuals who are often chosen as subjects for experiments because of the diminished risk involved. These experiments have their value, but they cannot be thought to represent the broad spectrum of the population adequately in their response to specific stressor agents.

HIGH-RISK GROUPS IN INDUSTRY

The identification of hypersusceptible individuals, especially in industry, has recently developed into a major issue between management and labor. Ashford (1976) has indicated that industrial physicians may continue the "blame the worker" attitude by management in their attempts to identify hypersusceptible workers. He suggests that numerous industrial physicians conceive of preventive medicine as identifying the high-risk individual and suggesting a job transfer for him or, with regard to preemployment examinations, for denial of a job offer. Ashford concluded that the goal of preventive medicine should be the removal of the hazardous conditions and not the hypersusceptible individual.

The protection of the high-risk individuals has also not been adequately considered by industrial health standards. "Threshold limit values" (TLVs) which have been developed for the past 30 years by the American Conference of Governmental Industrial Hygienists (ACGIH) and adopted as guidelines by the government are not designed to protect all workers (i.e., the hypersusceptibles). To date, TLVs have been derived for more than 450 chemical agents used in industry. In contrast to the intent of TLVs, the Occupational Safety and Health Act of 1970 demands that no worker will acquire any impairment of health (OSHAct, 1970). With the expected implementation of the OSHAct mandate to protect the health of all workers, the need for greater knowledge of hypersusceptible segments of the population becomes more urgent.

Table 26. The Identification and Quantification of High-Risk Groups

High-Risk Groups	Estimated Number of Individuals in United States Affected	Pollutant(s) to Which High-Risk Group Is (May Be) Hypersusceptible
	Developmental Processes	
Immature enzyme detoxification systems	Embryos, fetuses, and neonates to the age of approximately 2–3 months	pesticides PCBs
Immature immune system	Infants and children do not reach adult levels of IgA until the age of 10–12	respiratory irritants
Deficient immune system as a function of age	Progressive degeneration after adolescence	carcinogens respiratory irritants
Adolescents, especially carriers of porphyria genes	See population frequency for porphyrias (pp. 69–72)	chloroquine hexachlorobenzene lead variety of medical drugs
Differential absorption of pollutants as a function of age	Infants and young children	barium lead radium strontium
Retention of pollutants as a function of age	Individuals above the age of 50	fluoride
Pregnancy	Approximately 20 females per 50,000 females per year in the United States	anticholinesterase carbon monoxide insecticides lead
Circadian rhythms including phase shifts	All people have certain periods of the day when they are more susceptible to challenge	hydrocarbon carcinogens and probably most other pollutants
Infant stomach acidity	Infants	nitrates nitrites

Table 26 (Continued)

High-Risk Groups	Estimated Number of Individuals in United States Affected	Pollutant(s) to Which High-Risk Group Is (May Be) Hypersusceptible

Genetic Conditions

Albinism:		
Tyrosinase-positive	1 per 14,000 Negroes; 1 per 60,000 Caucasians; very high frequency in American Indians	ultraviolet radiation
Tyrosinase-negative	1 per 34,000 or 36,000 Negroes and Caucasians; 1 per 10,000 or 15,000 in Ireland	
Catalase deficiency:		ozone
Hypocatalasemia	5,000,000 heterozygotes	radiation
Acatalasemia	16,000 homozygotes	
Cholinesterase variants	Highly sensitive homozygous and heterozygous individuals of European ancestry have a combined frequency of about 1 per 1250; moderately sensitive genotypic variants of European ancestry have a frequency of 1 per 15,000	anticholinesterase insecticides
Crigler-Najjar syndrome	Few individuals live to adulthood	PCBs
Cystic fibrosis	8,000,000 heterozygous for CF, 100,000 homozygous for CF; most common among Caucasians of European ancestry	ozone respiratory irritants
Cystinosis	Most common among Caucasians of European ancestry; unknown frequency	cadmium lead mercury uranium
Cystinuria	1 per 200–250 individuals, although asymptomatic, are affected; 1 per 20,000–100,000 are homozygous	cadmium lead mercury uranium

Table 26 (Continued)

High-Risk Groups	Estimated Number of Individuals in United States Affected	Pollutant(s) to Which High-Risk Group Is (May Be) Hypersusceptible
Glucose-6-phosphate dehydrogenase deficiency (G-6-PD)	1,600,000 American black males (11 percent); Mediterranean Jews 11.0 percent, Greeks 1–2 percent, Sardinians 1–8 percent	carbon monoxide lead nitrate nitrite ozone radiation
Glutathione deficiency	Only a few cases have been reported	lead ozone
Glutathione peroxidase deficiency	Partial deficiency in some neonates; only a few adult cases reported	lead ozone
Glutathione reductase deficiency	Unknown, but thought to be rare	lead ozone
Gilbert's syndrome	6 percent of the normal, healthy adult population	PCBs
Immunoglobin A deficiency	500,000 homozygotes	respiratory irritants
Immunologic hypersensitivity	2 percent of some worker populations	isocyanates
Inducibility of aryl hydrocarbon hydroxylase	Approximately 45 percent of the general population are at high risk; 9 percent of the 45 percent are at very high risk	polycyclic aromatic hydrocarbons
Leber's optic atrophy	Thought to be rare	cyanide
Methemoglobin reductase deficiency	Found in many ethnic groups; gene frequency is uncertain but thought to be rare; most experience a temporary deficiency	nitrates nitrites ozone vanadium
Phenylketonuria	1 per 80 in the United States is a carrier; 1 per 25,600 has the disease	ultraviolet light

Table 26 (Continued)

High-Risk Groups	Estimated Number of Individuals in United States Affected	Pollutant(s) to Which High-Risk Group Is (May Be) Hypersusceptible
Porphyrias	1.5 per 100,000 in Sweden, Denmark, Ireland, West Australia; 3 per 1000 in South African whites; rare in blacks	chloroquine hexachlorobenzene lead various drugs including barbiturates, sulfonamides, and others
Serum alpha₁ antitrypsin	Approximately 4.0–9.0 percent individuals of northern European descent are heterozygotes; approximately 160,000 homozygotes, based on frequency found in Norway and Sweden	respiratory irritants smoking
Sickle cell		
Trait (heterozygote)	7–13 percent of American blacks, heterozygotes	aromatic amino and nitro compounds
Anemia (homozygote)	Homozygotic condition highly fatal	carbon monoxide cyanide
Sulfite oxidase	Unknown	sulfur dioxide sulfite
Thalassemia (Cooley's anemia)	Homozygous condition highly fatal; heterozygote frequency high (0.1–8.0 percent) in certain people of Italian, Greek, Syrian, and Negro origin	lead organic chemicals such as benzene and its derivatives ozone
Tyrosinemia	Frequency in general population unknown; frequency of heterozygous carriers in the Chicoutimi region of northern Quebec is carrier for every 20–31 people	cadmium lead mercury uranium
Wilson's disease	400,000 heterozygotes, 200 homozygotes; most frequent among Jews of Eastern European ancestry and non-Jews from the Mediterranean regions, especially Sicily	lead vanadium

Table 26 (Continued)

High-Risk Groups	Estimated Number of Individuals in United States Affected	Pollutant(s) to Which High-Risk Group Is (May Be) Hypersusceptible
Xeroderma pigmentosum	Unknown	ultraviolet radiation

Nutritional Deficiencies

Vitamin A	25 percent of children between 7 and 12 have lower than recommended dietary allowance (RDA), or slightly higher percent among the lower income groups	DDT hydrocarbon carcino- gens PCBs
Vitamin C	10–30 percent of infants, children and adults of low-low-income groups receive less than the RDA	arsenic cadmium carbon monoxide chromium DDT dieldrin lead mercury nitrites ozone
Vitamin E	7 percent of the general population are "physiologically deficient"	lead ozone
Calcium	65 percent of children between the ages of 2 and 3 receive less than the RDA	lead
Iron	98 percent of children between the ages of 2 and 3 receive less than the RDA	hydrocarbon carcinogens lead manganese
Magnesium	Most United States males have a partial deficiency	fluoride
Phosphorus	Deficiency in people with various kidney diseases	lead

Table 26 (Continued)

High-Risk Groups	Estimated Number of Individuals in United States Affected	Pollutant(s) to Which High-Risk Group Is (May Be) Hypersusceptible
Selenium	Unknown; deficiency thought to be rare	cadmium mercury ozone
Zinc	Unknown; deficiency in association with various diseases, but not thought to be widespread	cadmium
Riboflavin	30 percent of women and 10 percent of men aged 30–60 ingest less than two thirds of the RDA	hydrocarbon carcinogens lead ozone
Dietary protein	10 percent of women and 5 percent of men aged 30–60 ingest less than two thirds of the RDA for protein	DDT and other insecticides
Methionine	Unknown; but thought to be more frequent in individuals following certain vegetarian diets	DDT

Diseases

Kidney disease	Relates to genetic diseases (see cystinosis, cystinuria, tyrosinemia), bacterial and virus infections, and hypertensive disease	fluoride lead other heavy metals excessive sodium in diet
Liver disease	Relates to genetic diseases (Gilbert's syndrome) and virus infections	carbon tetrachloride DDT and other insecticides PCBs
Asthmatic diseases	4,000,000–10,000,000 of the general population	respiratory irritants: nitrogen dioxide ozone sulfates sulfur dioxide

Table 26 (Continued)

High-Risk Groups	Estimated Number of Individuals in United States Affected	Pollutant(s) to Which High-Risk Group Is (May Be) Hypersusceptible
Chronic respiratory disease	6,000,000–10,000,000 of the general population	respiratory irritants: nitrogen dioxide ozone sulphates sulfur dioxide
Heart disease	15,000,000 of the general population	cadmium carbon monoxide fluoride respiratory irritants: ozone sulfur dioxide sodium
Behavioral Activities		
Smoking	Widespread	cadmium hydrocarbons lead nickel radioactive compounds
Alcohol consumption	Widespread	lead pesticides PCBs
Drug taking	Widespread	pesticides PCBs numerous potential substances

ALTERNATE APPROACHES TO PROTECTING THOSE AT INCREASED RISK

Stewart (1976) feels that our laws should be changed so that we are not compelled to protect the most sensitive individual. He suggests that it is not realistic to impose an economic burden on the whole nation to protect the few. As an alternative, he recommended that these people should be given the chance to live in federally built and maintained artificial environments. This would be considerably cheaper than trying to maintain perfect air quality throughout the entire country. Of course, Stewart did not speculate on the financial cost of artificial environments, the problems of relocating people, and whether or not society as a whole would even minimally accept this as a viable alternative.

As unrealistic as the Stewart suggestion seems, it correctly implies that difficult decisions remain. It does seem economically, if not physically, impossible to remove all environmental stresses to no-effect levels. Consequently, we are forced to make cost-benefit analyses which ultimately end in value judgments and in the collective subjective interpretations of what is necessary for a reasonably satisfactory life style. The answers to such problems are not static but will be constantly modified as the society that makes them also evolves.

A more realistic alternative than the suggestion of Stewart can be seen in the behavior of asthmatics, individuals with tuberculosis, and others with clearly defined clinical diseases which entailed moving to another part of the country offering less respiratory stress. Of course, it should be mentioned that regional transport of pollutants makes this alternative not particularly favorable. Recent evidence by Rasmussen (1976) and Jeffries et al. (1976) that oxidants such as ozone may be transported from 100 to 1000 km supports this position. Also, the application of high-stack technology is known to influence the transportation of oxides of sulfur over a broad geographical area. Unfortunately, the regional transportation of pollutants seems to reduce the ability of any high-risk individual to "escape" the pollutant by moving from one part of the country to another. Yet, regional transportation is a rather new concept, and considerable uncertainty exists with respect to the magnitude of this phenomenon. It is certainly possible that regional transport of pollutants may turn out to be only a comparatively minor problem. Only future research will settle this issue. The idea of moving to a new and relatively unpolluted area would certainly be a valuable option, especially for those individuals in various high-risk groups. With this in mind, the idea of pollution-free zones (regions with negligible industrial-automobile pollution) for high-risk groups should be considered. Within this context, it is conceivable that permanent residents of such regions could only drive nonpolluting electric cars. Of course, considerable controversy surrounds interpretations of the Clean Air Act with regard to non-degradation of ambient air which is "cleaner" than that required by the national ambient air standards

(Bockrath, 1977). Realizing the legitimate right of industry to expand into new areas as well as the practical impossibility of reducing pollutants to safe levels for certain high-risk groups, it seems that specific pollution-free zones may offer a more realistic compromise to the current situation.

With regard to drinking water quality, it is generally accepted that there is an inverse relationship between the frequency of cardiovascular disease and water hardness. As early as 1963, an article in *Consumer Bulletin* had suggested that people with a family history of heart disease should consider moving to a hard-water area (see Figures 23 and 24 (pp. 131, 132), which show areas of hard and soft drinking water and rates of cardiovascular death). Many people try to circumvent these problems by buying bottled water, which is thought to contain no harmful substance, or by filtering their own home water supply.

TOXIC SUBSTANCES LEGISLATION

On October 11, 1976, President Ford signed into law the long awaited Toxic Substances Control Act (TSCA). The TSCA is hoped to provide the legal basis for the screening of potentially deleterious substances prior to their introduction into commerce. It is expected that the TSCA will fill a significant need in the existing environmental and health protection legislation. Previous laws, including the Clean Air Act, the Water Pollution Control Act, the Occupational Safety and Health Act, and the Consumer Product Safety Act, have been rather incomplete with regard to toxic substances regulation. More specifically, these laws initiate the regulation process after manufacture of the substances has begun, instead of beforehand. An exception to this general trend in environmental-occupational and consumer protection laws is the Federal Insecticide, Fungicide and Rodenticide Act (FIFRA) of 1972, which, similar to TSCA, provides what is called "front-end control" over pesticides (Carter, 1976).

It is generally recognized that this act represents the most advanced application of the concepts of preventive medicine in our public policy toward occupational and environmental health. This act should ultimately result in reducing by a considerable degree the flow of new toxic substances into the workplace. However, an almost completely overlooked component of TSCA, in complete contrast to laws such as the Clean Air Act, is the inclusion of a cost-benefit analysis associated with each substance considered. A practical consequence of the cost-benefit analysis is the recognition that a "safe" environment is not completely possible and that expectations for perfectly "clean" operations by industry should not necessarily be required. It also recognizes that certain products may be of such importance to our society that the adverse effects associated with their manufacture and distribution in the population may be outweighed by their benefits. The cost-benefit analysis with regard to industrial activities and the health of the environment represents a different attitude of the United States Congress in 1976 as compared with 1969, when the Clean Air Act was passed. The Congress of 1969 was probably not so aware as the 1976 Congress of the costs of environmental regulations to industry. The 1969 Congress was also probably not so aware of the concept of nonthreshold within a heterogeneous population. In effect, this new attitude provides an excellent approach to steering clear of potential hazards through its screening procedure but also recognizes that it is not possible to ensure the absolute protection of everyone.

The passage of TSCA also strongly implies the even greater need to focus considerable attention on the identification and quantification of high-risk groups. The expected health benefits to society from the passage of TSCA are directly related to our knowledge of high-risk groups. Without research into this area, the cost-benefit ratio will probably be more heavily weighted toward economic concerns as compared with health cost. Society needs a balance between risks and benefits. An inadequate data base for either side of the equation does not serve the best interests of our society.

SELECTING OF A POLLUTANT PRIORITY LIST

Every organization, including both federal and state environmental regulatory agencies, has budgetary limits and, therefore, must develop priority lists for prospective projects. Each region of the country has its own particular problems due to its unique geographical setting, the type of industry, or other reasons. For example, some cities have high SO_2 levels, but others have high ozone levels. Some regions may also have high levels of potentially toxic substances in the drinking water as a result of certain geological rock formations. However, despite the uniqueness of each region, there are some general considerations of high-risk populations which should be applied to the development of a pollutant ranking system in any region:

1. Identification of the actual pollutant level and the general toxicological properties of the pollutant. This includes a recognition of
   the pollutant levels that produce toxic responses in both the general
   public and the specific hypersusceptible segments of the population.
2. Characterization of the population with regard to age structure, sex
   ratios, racial background, nutritional and generalized health status
   (e.g., disease prevalence). This information is needed to determine
   the identification and quantification of high-risk groups. With such
   information, it is possible to determine the number of individuals
   who live in areas that have relatively high levels of pollutants to
   which they would be at high risk. A clear illustration of this situation
   is seen in cities that have a high ozone concentration and a large
   number of blacks with the G-6-PD deficiency as well as low levels of
   serum vitamin E. In this situation, a high level of environmental
   stressor (ozone) is in an area inhabited by those least capable of
   sustaining suitable homeostatic-compensatory responses to offset
   the stress. In most instances, this situation would be considered as
   high priority.
3. Consideration of the severity of the disease in priority determinations. The following questions should be investigated: Is the
   condition an aggravation of a preexisting disease (e.g., bronchitis); is
   the condition inherently of a self-limiting nature (hemolytic anemia
   associated with G-6-PD deficiency); or is the condition of a generally
   irreversible nature (cancer)? Higher priorities should be given to
   preventing diseases of an irreversible nature.

The establishment of pollutant priorities has recently become a
pressing issue in light of the TSCA. It is estimated that 1000 new
chemicals enter commerce each year. The EPA has publicly stated that
a priority list of the 50 most toxic substances of the 1000 new chemicals
manufactured each year will be given special attention with regard to
more detailed toxicological testing than is usually required (Olsen,
1976). How this list is chosen has not yet been defined. The health of
countless Americans in present and future generations depends on the
reliability of the pollutant priority selection process adopted by our
governmental decision makers.

Glossary

Acanthocytosis. A genetic disease which is characterized by malabsorption of fat and leads to degenerative changes in the nervous system.

Acrolein. A volatile acrid liquid, $CH_2:CHCHO$.

Aerosol. A colloid system in which the continuous phase (dispersion medium) is a gas, e.g., fog.

Aliphatic. A term applied to the "open chain" or fatty acid series of hydrocarbons.

Alkalosis. A pathologic condition resulting from accumulation of base in, or loss of acid from, the body.

Alkyl. The radical which results when an aliphatic hydrocarbon loses one hydrogen atom.

Alkylate. To treat with an alkylating agent.

Alkylation. The substitution of an alkyl group for an active hydrogen atom in an organic compound.

Allograft. A graft of tissue between individuals of the same species but of a different genotype.

Alpha Radiation. High-speed helium nuclei which have been ejected from radioactive substances.

Anaphylaxis. An unusual or exaggerated allergic reaction of an organism to foreign protein or other substances.

Androgen. Any substance that possesses masculinizing activities, such as the testicular hormone.

Ankylosis. Immobility and consolidation of a joint.

Anorexia. Lack or loss of the appetite for food.

Antibody. A specific protein produced by an organism which can combine specifically with its antigen.

Antigen. A substance which, when introduced into the body of an organism, stimulates the production of antibodies; usually a protein or carbohydrate.

Aromatic.   Denoting a compound characterized by the benzene ring.

Arteriosclerosis.   A group of diseases characterized by thickening and loss of elasticity of arterial walls.

Atherosclerosis.   An extremely common form of arteriosclerosis in which deposits of yellowish plaques (atheromas) containing cholesterol, lipoid material, and lipophages are formed within the intima and inner media of large- and medium-sized arteries.

Audiogenic.   Produced by sound.

Autoimmunity.   A condition characterized by a specific immune response against the constituents of the body's own tissue.

Autosomes.   Chromosomes other than sex chromosomes.

Bacteriophage.   A virus that lyses bacteria.

Beta Radiation.   Electrons ejected from radioactive substances.

Biliary.   Pertaining to the bile, to the bile ducts, or to the gallbladder.

Bilirubin.   A bile pigment: it is a breakdown product of heme mainly formed from the degradation of erythrocyte (red blood cell) hemoglobin but also formed by breakdown of other heme pigments, e.g., cytochromes. Bilirubin normally circulates in plasma as a complex with albumin and is taken up by the liver cells and conjugated to form bilirubin diglucuronide, which is the water-soluble pigment excreted in the urine.

Bursa of Fabricius.   An epithelial outgrowth of the cloaca in chick embryos which develops in a manner similar to that of the thymus in mammals, atrophying after 5 or 6 months and persisting as a fibrous remnant in sexually mature birds. It contains lymphoid follicles and before involution is a site of formation of lymphocytes associated with humoral immunity.

Calcinosis.   A condition marked by the deposition of calcium salts in various tissues of the body.

Calculus, Calculi (pl.).   An abnormal pebble or stone occurring within the animal body and usually composed of mineral salts.

Carcinoma.   A malignant new growth made up of epithelial cells tending to infiltrate the surrounding tissues and giving rise to metastases.

Catabolism.   Any destructive process by which complex substances are converted by living cells into more simple compounds.

Cell-Mediated Immunity.   Specific immunity which is mediated by small lymphocytes and is dependent on the presence of the thymus at birth.

Ceruloplasmin. An alpha$_2$-globulin of the plasma, being a glycoprotein in which approximately 96 percent of plasma copper is transported.

Chylomicron. A stable droplet containing 86 percent triglyceride fat, 3 percent cholesterol, 9 percent phospholipids, and 2 percent protein, found in the intestinal lymphatics and blood during and after meals; this is the form in which absorbed long-chain fats and cholesterol are transported from the intestine.

Cirrhosis. A liver disease characterized pathologically by loss of the normal microscopic lobular architecture, with fibrosis and nodular regeneration.

Coenzyme. The nonprotein part of an enzyme.

Colostrum. The thin, yellow, milky fluid secreted by the mammary gland a few days before or after birth.

Congenital. Existing at, and usually before, birth; referring to conditions that are present at birth, regardless of origin.

Corticosteroid. Any of the steroids produced by the adrenal cortex, including costisol, corticosterone, aldosterone, etc.

Cyanosis. A bluish discoloration, applied especially to such discoloration of skin and mucous membranes due to excessive concentration of methemoglobin in the blood.

Cystine. An amino acid, $(S \cdot CH_2CH(NH_2)-COOH)_2$, produced by the digestion of proteins. Cystine is the chief sulfur-containing compound of the protein molecule and is readily reduced to two molecules of cysteine.

Cytochrome. Any of a class of hemoproteins whose principal biologic function is electron transport by virtue of a reversible valency change of its heme iron.

Degranulation. The process of losing granules; said of granular cells.

Diphenylhydantoin. Chemical name: 5-5-diphenyl-2,4-imidazolidin-edione; $(C_{15}H_{12}N_2O_2)$; used as an anticonvulsant in grand mal epilepsy.

Dyskinesia. Impairment of the ability to move.

Dysmenorrhea. Painful menstruation.

Electron Transport System. A structually and functionally organized system which generates ATP in cells.

Emphysema (Pulmonary). A condition of the lung characterized by increase beyond normal in the size of air spaces distal to the terminal bronchioles, either from dilatation of the alveoli or from destruction of their walls.

Endonuclease. A nuclease that cleaves internal bonds of polynucleotides.

Endoplasmic Reticulum. A system of vesicles or canaliculi in the cytoplasm of cells as revealed by the electron microscope.

Endotoxin. A heat-stable toxin present in the bacterial cell but not in cell-free filtrates of cultures of intact bacteria.

Epoxidation. The process by which a molecule containing one atom of oxygen bound to two different carbon atoms is formed.

Erythropoiesis. The production of erythrocytes.

Exocrine. Secreting outwardly via a duct.

Fibroblast. A connective tissue cell.

Flavoprotein. A conjugated protein in which the prosthetic group contains a flavin.

Galvanized Pipes. Pipes coated with zinc; since zinc is often found with cadmium, the pipes are also coated inadvertently with cadmium.

Globulin. Type of protein which is insoluble or sparingly soluble in water.

Glucoside. A glycoside in which the sugar constituent is glucose.

Glucuronide. Any compound with glucuronic acid.

Glycogen. The chief carbohydrate storage material in animals.

Glycolysis. The breaking down of sugars into simple compounds, chiefly pyruvate or lactate.

Glycosuria. The presence of an abnormal amount of glucose in the urine.

Hardness of Water. A quality of water produced by soluble salts of calcium and magnesium or other substances which form an insoluble curd with soap and thus interfere with its cleaning power.

Heart Failure (Congestive). A clinical syndrome due to heart disease and characterized by breathlessness and abnormal sodium and water retention, resulting in edema; the ingestion may occur in the lungs, or in the peripheral circulation, or in both, depending on whether the failure is right-sided, left-sided, or general.

Hemolysis. The separation of the hemoglobin from the red cells and its appearance in the plasma.

Hepatocyte: A particular type of liver cell.

Heterozygote, Heterozygous (adj.). An organism with genetic alleles unlike members of any given pair or series of alleles, that consequently produces unlike gametes.

Histamine. An amine, beta-imidazolylethylamine, $CH \cdot NH \cdot CH \cdot N \cdot C \cdot$

$CH_2 \cdot NH_2$, formed by decarboxylation of histidine and occurring in many animal and vegetable tissues, particularly in the granules of mast cells and basophils. It has at least three important functions: (1) dilation of capillaries, which increases capillary permeability and results in a drop of blood pressure, (2) constriction of the bronchial smooth muscle of the lungs, and (3) induction of increased gastric secretion. It is also implicated as a mediator of immediate hypersensitivity.

Homozygote, Homozygous (adj.).   An organism whose chromosomes carry identical members of any given pair of alleles. The gametes are therefore all alike with respect to this locus.

Hydrazine.   A colorless gaseous diamine, $H_2N \cdot NH_2$; also any member of a group of its substitution derivatives.

Hypernitraemia.   Excessive nitrogen in the blood.

Hypertension.   Persistently high arterial blood pressure.

Hypophosphatemia.   An abnormally decreased amount of phosphates in the blood; it may be found in hyperparathyroidism, rickets, osteomalacia, and several renal tubular abnormalities, including the Fanconi syndrome.

Hypoxia.   Low oxygen content or tension.

Icterus.   Jaundice.

Immunogen.   Any substance that is capable of eliciting an immune response.

Immunoglobulin.   A protein of animal origin endowed with known antibody activity. Immunoglobulins function as specific antibodies and are responsible for the humoral aspects of immunity. There are five antigenically different kinds of immunoglobulins, listed as IgA, IgD, IgE, IgM, and IgG.

Infarction.   The formation of an area of coagulation necrosis (infarct) in a tissue due to local deficiency of blood resulting from obstruction of circulation to the area.

In Situ.   Confined to the site of origin without invasion of neighboring tissues.

In Vitro.   Within a glass; observable in a test tube; in an artificial environment.

In Vivo.   Within the living body.

Ionizing Radiation.   High-energy radiation which interacts to provide ion pairs in matter.

Keratin.   A scleroprotein which is the principal constituent of epidermis, hair, nails, horny tissues, and the organic matrix of the enamel of the teeth.

Ketone. A compound containing $\overset{O}{\underset{\parallel}{-C-}}$ attached to two carbon atoms, for example,
$$CH_3-\overset{O}{\underset{\parallel}{C}}-CH_3$$

Krebs Cycle. The tricarboxylic acid cycle; the cyclic metabolic mechanism by which the complete oxidation of the acetyl-coenzyme A is effected.

Lavage. The washing out of an organ.

Leukocyte. A generalized term for white blood cell.

Lymph. A transparent, slightly yellow liquid of alkaline reaction, found in the lymphatic vessels and derived from the tissue fluids; it consists of a liquid portion and of cells, most of which are lymphocytes. Lymph is collected from all parts of the body and returned to the blood via the lymphatic system.

Lymph Node. Any of the accumulations of lymphoid tissue organized or definite lymphoid organs situated along the course of lymphatic vessels; the lymph nodes are the main source of lymphocytes of the peripheral blood.

Lymphocyte. Type of nongranular nonphagocytic white blood cell.

Macrophage. Any of the large, highly phagocytic cells with a small, oval, sometimes indented nucleus and inconspicuous nucleoli, occurring in the walls of blood vessels and in loose connective tissue. They are components of the reticuloendothelial system and are usually immobile but when stimulated by inflammation become actively mobile.

Ménière's Disease (Syndrome). Deafness, tinnitus, vertigo resulting from nonsuppurative disease of the labyrinth.

Menorrhagia. Excessive uterine bleeding occurring at the regular intervals of menstruation, the period of flow being of greater than usual duration.

Metachromasia. A condition in which tissues do not stain true with a given stain.

Methemoglobin. A compound formed from hemoglobin by oxidation of the ferrous to the ferric state; it is unable to bind oxygen reversibly.

Mitochrondria. Small, spherical to rod-shaped organelles found in the cytoplasm of cells, enclosed in a double membrane, the inner one having infoldings called cristae. They are the principal sites of the generation of energy resulting from the oxidation of foodstuffs and they contain the enzymes of the Krebs and fatty acid cycles and the respiratory pathways.

Mitogen. A substance which induces mitosis.

Monocyte. A mononuclear phagocytic leukocyte; formerly called large mononuclear leukocyte.

Myocardial Infarction. Gross necrosis of the myocardium, as a result of interruption of the blood supply to the area.

Myocardium. The middle and thickest layer of the heart wall, composed of cardiac muscle.

Nephritis. Inflammation of the kidney; a focal or diffuse proliferative or destructive process which may involve the glomerulus, tubule, or interstitial renal tissue.

Nephropathy. Disease of kidneys.

Nephrosis. Any disease of the kidney, especially any disease of the kidneys characterized by purely degenerative lesions of the renal tubules—as opposed to nephritis—and marked by edema, albuminuria, and decreased serum albumin.

Opthalmalogic. Pertaining to the eye.

Papilloma. A branching or lobulated benign tumor derived from epithelium.

Paraquat. A poisonous dipyridilium compound used as a weed killer.

Peroxide. The oxide of any element which contains more oxygen than any other. More correctly applied to compounds having such linkage as —O—O—; for example, hydrogen peroxide, H—O—O—H.

Phenol. A generic term for any organic compound containing one or more hydroxyl groups attached to an aromatic or carbon ring.

Phenotype. The appearance of an individual in contrast to its genetic constitution.

Plasma. The fluid portion of the blood in which the corpuscles are suspended. Plasma is to be distinguished from serum, which is the all-free portion of the blood from which the fibrinogen has been separated in the process of clotting.

Plasma Cell. A spherical or ellipsoidal cell, with a single, eccentrically placed nucleus containing clumped chromatin, an area of perinuclear clearing, and generally abundant, sometimes vacuoled cytoplasm; plasma cells are functionally involved, under various circumstances, in the synthesis of immunoglobulins and their components.

Plumbism. Lead poisoning.

Plummer-Vinson Syndrome. Dysphagia with glossitis, hypochromic anemia, spenomegaly, and atrophy in the mouth, pharynx, and upper end of the esophagus.

Protease. A general term for a proteolytic enzyme.

Quinone. Any benzene derivative in which two hydrogen atoms are replaced by two oxygen atoms.

Reticulocyte. A young red blood cell showing a basophilic reticulum under vital staining.

Rickets. A condition caused by deficiency of vitamin D, especially in infancy and childhood, with disturbance of normal bone formation. *Vitamin D–resistant* rickets—a condition almost indistinguishable from ordinary rickets clinically but resistant to unusually large doses of vitamin D.

Scotoma Scotomata (pl.). An area of depressed vision within the visual field, surrounded by an area of less depressed or of normal vision.

Sex-Linked Genetic Trait. Characteristics due to genes located on the x-chromosomes.

Spondylitis. Inflammation of the vertebrae.

Sulfite. Any salt of sulfurous acid.

T-Cell (Thymus-Dependent Cell). Depending on passage through or on being influenced by the thymus for the development of function.

Tetany. A syndrome manifested by sharp flexion of the wrist and ankle joints, muscle twitching, cramps, and convulsions. It is due to abnormal calcium metabolism.

Thiocyanate. A salt analogous in composition to a cyanate but containing sulfur instead of oxygen.

Thymectomy. The removal of the thymus.

Thymus. A ductless, glandlike body situated in the chest region which reaches into maximum development during the early years of childhood and then declines in size. It is known to play a significant role in the body's immune responses.

Tradescantia. Any plants of a genus (*Tradescantia*), family (Commelinaceae) of American herbs; the spiderwort.

Tyrosyluria. The increased urinary excretion of parahydroxyphenyl compounds derived from tyrosine, as in tyrosinemia.

Uremia. The retention of excessive by-products of protein metabolism in the blood, and the toxic condition produced thereby. It is marked by nausea, vomiting, headache, vertigo, dimness of vision, coma, or convulsions. It is due to nephron function inadequate to excrete urea and other products of protein metabolism.

Vertigo. Dizziness; giddiness.

References

Aaronson (Editorial) (1971). "Spectrum." *Environment*, June, p. 26.

Abe, M. (1967). Effects of mixed NO_2–SO_2 gas on human pulmonary functions. *Bull. Tokyo Med. Dent. Univ.*, **14**: 415–433.

Abt, A. F. (1942). The human skin as an indicator of the detoxifying action of vitamin C (ascorbic acid) in reactions due to arsenicals used in antisyphilitic therapy. *U.S. Navy Bull.*, **40**: 291–303.

Ackermann, E. (1971). Die Michaelis-Menten Konstant und Maximalen Reaktironsgesch windigkeiten einiger Pharmaka in den Lebernikrosomen des Mesenchen verglichen mit der Ratte beiderlea Geschlechts. *Biochem. Pharmacol.*, **20**: 1920.

Ackermann, E., and Rane, A. (1971). The monooxygenase system in the human fetal liver: Subcellular distribution and studies on *in vitro* metabolism of aniline. *Chemico-Biol. Interactions*, **3**: 233.

Ackner, B., Cooper, J. E., Gray, C. H., Kelly, M., and Nicholson, D. C. (1961). Excretion of porphobilinogen and δ-aminolaevulinic acid in acute porphyria. *Lancet*, **1**: 1256.

Adamson, R. H., and Davies, D. S. (1973). Comparative aspects of absorption, distribution, metabolism and excretion of drugs. Comp. Pharmacol., **2**: 851.

Aebi, H. (1967). The investigation of inherited enzyme deficiencies with special reference to acatalasia. *Proc. 3rd Int. Congr. Hum. Genet.*, Chicago, 1966, J. F. Crow and J. V. Neel (Eds.), Johns Hopkins, Baltimore, p. 189.

Aebi, H., Baggiolini, M., Dewald, B., Lauber, E., Suter, H., Micheli, A., Frei, J., and Marti, H. R. (1962/3). Observations in two Swiss families with acatalasia II. *Enzymol. Biol. Clin.*, **2**: 1.

Aebi, H., Heiniger, J. P., Butler, R., and Hassig, A. (1961). Two cases of acatalasia in Switzerland. *Experientia*, **17**: 466.

Aebi, H., Heiniger, J. P., and Suter, H. (1962). Methaemoglobinbildung dürch Röntgenbestrahlen in Normalen and Ratalasefreien erthrocyten des menschen. *Experientia*, **18**: 129.

Aebi, H., and Suter, H. (1969). Catalase. In *Biochemical Methods in Red Cell Genetics*, J. J. Yunis (Ed.), Academic, New York, p. 255.

Aebi, H., and Suter, H. (1972). Acatalasemia. In *The Metabolic Basis of Inherited Disease*, 3rd ed., J. B. Stanbury, J. B. Wyngaarden, and D. S. Fredrickson (Eds.), McGraw-Hill, New York, pp. 1710–1729.

Alarie, Y. (1976). Rebutal. *Arch. Environ. Health*, March/April, pp. 110–112.

Alexanderson, B., Price Evans, D. A., and Sjoquist, F. (1969). Steady-state plasma levels of nortriptyline in twins: Influence of genetic factors and drug therapy. *Br. Med. J.*, **4**: 764–768.

Allaway, W. H., Kubota, J., Losee, F., and Roth, M. (1968). Selenium, molybdenum, and vanadium in human blood. *Arch. Eniron. Health*, **16**: 342.

Allen, J. R. (1975). Response of the non-human primate to polychlorinated biphenyl exposure. *Fed. Proc.*, **34**(8): 1675.

Allison, A. C. (1956). Observations on the sickling phenomenon and on the distribution of different hemoglobin types in erythrocyte populations. *Clin. Sci.*, **15**: 497.

Allison, S. D., and Wong, K. L. (1967). Skin cancer: Some ethnic differences. In *Environments of Man*, J. B. Bresler (Ed.), Addison–Wesley, Reading, Mass., pp. 67–71.

Altshuller, A. P. (1975). Evaluation of oxidant results at camp sites in the United States. *J. Air Pollut. Control*, **25**: 19.

Alwall, N., Laurell, C. B., and Nilsby, I. (1946). Studies on heredity in cases of "Non-hemolytic bilirubinemia without direct Van den Bergh reaction" (hereditary, non-hemolytic bilirubinemia). *Acta Med. Scand.*, **124**: 114.

Ammann, A. J., Roth, J., and Hong, R. (1970). Recurrent sinopulmonary infections, mental retardation and combined IgA and IgE deficiency. *J. Pediat.*, **77**: 802–804.

Ammann, P., Herwig, K., and Baumann, T. (1968). Vitamin A excess. *Helv. Paediatr. Acta*, **23**: 137.

Anchev, N., Popov, I., and Ikonopisov, R. L. (1966). Epidemiology of malignant melanoma in Bulgaria. In *Structure and Control of the Melanocyte*, G. Della Porta, and O. Muhlbock (Eds.), Springer–Verlag, New York, pp. 286–291.

Anderson, C. M., Freeman, M., Allan, J., and Hubbard, L. (1962). Observations on (i) sweat sodium levels in relation to chronic respiratory disease in adults and (ii) the incidence of respiratory and other disease in parents and siblings of patients with fibrocystic disease of the pancreas. *Med. J. Austr.*, **1**: 965.

Anderson, E. W., Andelman, R. J., Stravch, J. M., et al. (1973). Effect of low level carbon monoxide exposure on onset and duration of angina pectoris. *Ann. Intern. Med.*, **79**: 46–50.

Anderson, T. W., LeRiche, W. H., and Mackay, J. S. (1969). Sudden death and ischemic heart diseases: Correlation with hardness and local water supply. *New Eng. J. Med.*, **280**: 805.

Andrzejewski, S. A. (1966). Studies on the toxicity of tobacco and tobacco smoke. *Acta Med. Pol.*, **5**: 407–408.

Anonymous. (1968). Screening for inborn errors of metabolism. *WHO Tech. Rep. Ser.*, 401.

Anonymous. (1969). Chronic cyanide neurotoxicity. *Lancet*, Nov. 3, pp. 942–943.

Anonymous. (Jan. 21, 1971). Nitrates, nitrites, and nitrosamines. *FDA Fact Sheet.*

Anonymous. (Winter, 1971). Children lack vitamin A. *J. Nutr. Educ.*, p. 82.

Anonymous. (1973). Pharmacogenetics. *WHO Tech. Rep. Ser.*, 524.

Anthony, D. D., Schaffner, F., Popper, H., and Hutterer, F. (1962). Effect of cortisone on hepatic structure during subacute ethionine intoxication in the rat: An electron microscopic study. *Mol. Pathol.*, **1**: 113–121.

Arias, I. M. (1962). Chronic unconjugated hyperbilirubinemia without overt signs of hemolysis in adolescents and adults. *J. Clin. Invest.*, **41**: 2233.

Arias, I. M., Gartner, L. M., Cohen, M., Ezzer, J. B., and Levi, A. J. (1969). Chronic nonhemolytic unconjugated hyperbilirubinemia with glucuronyl transferase deficiency. *Am. J. Med.*, **47**: 395.

Ariens, E. J., Simonis, A. M., and Offermeir, J. (1976). *Introduction to General Toxicology*, Academic, New York, pp. 44–87.

Aronow, W. S., et al. (1972). Effect of freeway travel on angina pectoris. *Ann. Intern. Med.*, **77**: 669–676.

Aronow, W. S., and Isbell, M. W. (1973). Carbon monoxide effect on exercise-induced angina pectoris. *Ann. Intern. Med.*, **79**: 392–395.

Ashford, N. A. (1976). *Crisis in the Workplace: Occupational Disease and Injury*, M.I.T., Cambridge, Mass., pp. 66–92.

Ayers, S. M., Giannelli, S., Jr., and Mueller, H. (1970). Myocardial and systemic responses to carboxyhemoglobin. *Ann. N.Y. Acad. Sci.*, **174**: 268–293.

Bachmann, R. (1965). Studies on the serum gamma-A-globulin level: III. The frequency of A-gamma A-globulinemia. *Scand. J. Clin. Lab. Invest.*, **17**: 316–320.

Baernstein, H. D., and Grand, J. A. (1942). The relation of protein intake to lead poisoning in rats. *J. Pharmacol. Exp. Ther.*, **74**: 18.

Barnes, J. M., and Davies, D. R. (1951). Blood cholinesterase levels in workers exposed to organo-phosphorous insecticides. *Br. Med. J.*, **2**: 816.

Barnes, M. G., Komarmy, L., and Novack, A. H. (1972). A comprehensive screening program for hemoglobinopathies. *J. Am. Med. Assoc.*, **219**: 701.

Barron, E. S. G., and Dickman, S. (1949). Studies on the mechanism of action of ionizing radiation: II. Inhibition of sulfhydryl enzymes by alpha, beta and gamma rays. *J. Gen. Physiol.*, **32**: 595.

Bates, D. V. (1972). Air pollutants and the human lung. *Am. Rev. Respir. Dis.*, **105**: 1–13.

Batten, J., Muir, D., Simon, G., and Carter, C. (1963). The prevalence of respiratory disease in heterozygotes for the gene for fibrocystic disease of the pancreas. *Lancet*, **1**: 1348.

Battigelli, M. C., and Gamble, J. F. (1976). From sulfur to sulfate: Ancient and recent considerations. *J. Occup. Med.*, **18**(5): 334–341.

Bearn, A. G. (1953). Genetic and biochemical aspects of Wilson's disease. *Am. J. Med.*, **15**: 442.

Bearn, A. G. (1972). Wilson's disease. In *Metabolic Basis of Inherited Disease*, 3rd ed., J. B. Stanbury, J. B. Wyngaarden, and D. S. Fredrickson (Eds.), McGraw-Hill, New York, p. 1033.

Bearn, A. G., Yu, T. F., and Guttman, A. B. (1957). Renal function in Wilson's disease. *J. Clin. Invest.*, **36**: 1107.

Beckman, R. (1955). Vitamin E physiology and clinical significance. *Z. Vitam. Horm. Fermentforsch.*, **7**: 153, 281.

Belanger, L. F., Visek, W. J., Lotz, W. E., and Comar, C. L. (1958). Rachitomimetic effects of fluoride feeding on the skeletal tissues of growing pigs. *Am. J. Pathol.*, **34**: 25.

Benitz, K. F., and Diermeier, H. F. (1964). Renal toxicity of tetracycline degradation products. *Proc. Soc. Exp. Biol. Med.*, **115**: 930.

Bennet, R. A., and Whigham, A. (1964). Chloroform sensitivity of mice. *Nature* (London), **204**: 1328.

Berg, J. W. (1975). Diet. In *Persons at High Risk of Cancer: An Approach to Cancer Etiology and Control*, J. F. Fraumeni, Jr. (Ed.), Academic, New York, pp. 201–224.

Berg, J. W., Haenszel, W., and Devesa, S. S. (1973). Epidemiology of gastrointestinal cancer. *Seventh Nat. Cancer Proc.*, pp. 459–464.

Berglund, F. (1967). Levels of polychlorinated biphenyl in food in Sweden.*Environ. Health Perspect.*, **1**: 67.

Bernstein, R. E. (1962). A rapid screening dye test for the detection of glucose-6-phosphate dehydrogenase deficiency in red cells. *Nature*, **194**: 192.

Beutler, E. (1957). The glutathione instability of drug-sensitive red cells: A new method for the *in vitro* detection of drug sensitivity. *J. Lab. Clin. Med.*, **49**: 84.

Beutler, E. (1959). The hemolytic effect of primaquine and related compounds: A review. *Blood*, **14**: 103.

Beutler, E. (1969). Glutathione reductase: Stimulation in normal subjects by riboflavin supplementation. *Science*, **165**: 613–615.

Beutler, E. (1971). *Red Cell Metabolism: A Manual of Biochemical Methods*, Grune and Stratton, New York.

Beutler, E. (1972). Glucose-6-phosphate dehydrogenese deficiency. In *The Metabolic Basis of Inherited Disease*, 3rd ed., J. B. Stanbury, J. B. Wyngaarden, and D. S. Fredrickson (Eds.), McGraw-Hill, New York, pp. 1358–1388.

Beutler, E. (1973). Screening for glucose-6-phosphate dehydrogenase deficiency. In *Genetic Polymorphisms and Diseases in Man*, Academic, New York, p. 224.

Beutler, E., Dern, R. J., and Alving, A. S. (1955). The hemolytic effect of primaquine: VI. An *in vitro* test for sensitivity of erythrocytes to primaquine. *J. Lab. Clin. Med.*, **45**: 40.

Beutler, E., Dern, R. J., Flanagan, C. L., and Alving, A. S. (1955). The hemolytic effect of primaquine: VII. Biochemical studies of studies of drug-sensitive erythrocytes. *J. Lab. Clin. Med.*, **45**: 286.

Beutler, E., and Mitchell, M. (1968). Special modifications of the fluorescent screening method for glucose-6-phosphate dehydrogenase deficiency. *Blood*, **32**: 816.

Bianchi, G., Fox, V., Difrancesco, G. F., Giovannetti, A. M., and Pagetti, D. (1974). Blood pressure changes produced by kidney cross-transplantation between spontaneously hypertensive rats (SHR) and normotensive rats (NR). *Clin. Sci. Mol. Med.*, **47**: 435.

Bickel, H., and Souchon, F. (1955). Die papierchromatographie in der Kinderheilkunde. *Arch. Kinderh.*, Suppl., **31**: 101.

Biersteken, K. (1967). Drinkwaterzachtheid en sterfte. *Tijdschr. Soc. Geneeskd.*, **45**: 658.

Billing, B. H. (1970). Bilirubin metabolism and jaundice with special reference to unconjugated hyperbilirubinemia. *Ann. Clin. Biochem.*, **7**: 69–74.

Bjernulf, A., Johnsson, S. G. O., and Parrow, A. (1971). Immunoglobulin studies in gastrointestinal dysfunction with special reference to IgA deficiency. *Acta Med. Scand.*, **190**: 71–77.

Block, W. D., and Cornish, H. H. (1959). Metabolism of biphenyl and 4-chlorobiphenyl in the rabbit. *J. Biol. Chem.*, **234**(12): 3301.

Blum, H. F. (1948). Sunlight as a causal factor in cancer of the skin of man. *J. Nat. Cancer Inst.*, **9**: 247–258.

Blum, H. F. (1959). *Carcinogenesis by Ultraviolet Light*, Princeton University Press, Princeton, N.J.

Bockrath, J. (1977). *Environmental Law for Engineers, Scientists, and Managers*, McGraw-Hill, New York, pp. 241–246.

Bodansky, O. (1951). Methemoglobin and methemoglobin-producing compounds. *Pharmacol. Rev.*, **3**: 144.

Bodegard, G., Gentz, J., Lindblad, B., Lindstedt, S., and Zetterstrom, R. (1969). Hereditary tyrosinemia: III. On the differential diagnosis and the lack of effect of early dietary treatment. *Acta Paediatr. Scand.*, **58**: 37.

Bogert, L. J., Briggs, G. M., and Calloway, D. H. (1973). *Nutrition and Physical Fitness*, 9th ed., Saunders, Philadelphia, p. 178.

Borg, D. C., and Cotzias, G. C. (1958). Incorporation of manganese into erythrocytes as evidence for a manganese porphyrin in man. *Nature*, **182**: 1677.

Borst, J. G. G., and Borst, DeGa. (1963). Hypertension explained by Starling's theory of circulatory homeostasis. *Lancet*, **1**: 677.

Boulos, B. (1976). School of Public Health, University of Illinois (personal communication).

Bourne, J. G., Collier, H. O., and Somers, G. F. (1952). Succinylcholine (succinoylcholine): Muscle relaxant of short action. *Lancet*, **1**: 1225.

Bourquin, A., and Musmanno, E. (1953). Effect of smoking on the ascorbic acid content of whole blood. *Am. J. Dig. Dis.*, **20**: 75–77.

Boutin, D., and Brodeur, J. (1969). An automated method for the determination of pseudocholinesterase variants. *Clin. Biochem.*, **2**: 187.

Bowman, J. E., Carson, P. E., Frischer, H., Kahn, M., and Ajmar, F. A. (1964). A capillary tube, nile blue methemoglobin reduction test for glucose-6-phosphate dehydrogenase deficiency. *Proc. X Int. Congr. Blood Transfus.*, Stockholm, p. 592.

Boyd, E. M. (1969). Dietary protein and pesticide toxicity in male weanling rats. *Bull. WHO*, **40**: 801.

Brand, E., Harris, M. M., and Biloon, S. (1930). Cystinuria: Excretion of a free cystine. *J. Biol. Chem.*, **86**: 315.

Brandtzaeg, P. (1971). Human secretion from individuals with selective secretion of defective synthesis of serum immunoglobulins. *Clin. Exp. Immunol.*, **8**: 69–85.

Brandtzaeg, P., Fjellander, I., and Gjeruldsen, S. T. (1968). Immunoglobulin M: Local synthesis and selective secretion in patients with immunoglobulin A deficiency. *Science*, **160**: 789–791.

Brewer, G. J., Tarlov, A. R., and Alving, A. S. (1960). Methaemoglobin reduction test: A new, simple *in vitro* test for identifying primaquine-sensitivity. *Bull. WHO*, **22**: 633.

Briscoe, W. A., Kueppers, F., Davis, A. L., and Bearn, A. G. (1966). A case of inherited deficiency of serum alpha₁-antitrypsin associated with pulmonary emphysema. *Am. Rev. Respir. Dis.*, **94**: 529–539.

Brodie, B. B., Gillette, J. R., and LaDu, B. N. (1958). Enzymatic metabolism of drugs and other foreign compounds. *Ann. Rev. Biochem.*, **27**: 427, 455.

Brook, M., and Grimshaw, J. J. (1968). Vitamin C concentration of plasma and leucocytes as related to smoking habits, age and sex of humans. *Am. J. Clin. Nutr.*, **21**: 1254–1258.

Brown, D. R., McGandy, R. B., Gallie, E., and Doyle, J. T. (1958). Magnesium lipid relations in health and in patients with myocardial infarction. *Lancet*, **2**: 933.

Buckley, R. D., Hackney, J. D., Clark, K., and Posin, C. (1975). Ozone and human blood. *Arch. Environ. Health*, **30**: 40.

Burns, J. J., Conney, A. H., and Koster, R. (1963). Stimulatory effect of chronic administration on drug-metabolizing enzymes in liver microsomes. *Ann. N.Y. Acad. Sci.*, **104**: 881–893.

Burrell, R. J., Roach, W. A., and Shadwell, A. (1966). Esophageal cancer in the Bantu of the Transkei associated with mineral deficiency in garden plants. *J. Natl. Cancer Inst.*, **36**: 201–214.

Burse, V. W., Kimbrough, R. D., Villanueva, E. C., Jennings, R. W., Linder, R. E., and Sovocovl, G. W. (1974). Polychlorinated biphenyls—storage, distribution, excretion and recovery: Liver morphology after prolonged dietary ingestion. *Arch. Environ. Health*, **29**: 301.

Calabrese, E. J. (1977). Insufficient conjugate glucuronidation: A possible cause of PCB toxicity. *Med. Hypothesis*, **3** (4): 162–165.

Calabrese, E. J. (1978a). Environmental quality indices predicted by evolutionary theory. *Med. Hypotheses* (in press).

Calabrese, E. J., Kajola, W., and Carnow, B. W. (1977). Ozone: A possible cause of hemolytic anemia in glucose-6-phosphate dehydrogenase deficient individuals. *J. Toxicol. Environ. Health*, **2**: 709–712.

Calabrese, E. J., and Sorensen, A. (1978). The health implications of PCBs with particular emphasis on high risk groups. *Rev. on Environ. Health* (in press).

Calder, J. H., Curtis, R. C., and Fore, H. (1963). Comparison of the vitamin C in plasma and leukocytes of smokers and non-smokers. *Lancet*, **1**: 556.

Cam, C., and Nigogoysan, G. (1963). Acquired toxic porphyria cutanea tarda due to hexachlorobenzene. *J. Am. Med. Assoc.*, **183**: 88.

Carnow, B. W. (1966). Air pollution and respiratory diseases. *Sci. Citizen*, May, p. 1.

Carnow, B. W. (1970). Relationship of SO_2 levels to morbidity and mortality in (high risk) populations. *Air Pollut. Med. Res. Conf.*, Oct. 5, 1970, New Orleans, La.

Carnow, B. W., and Carnow, V. (1974). Air pollution, morbidity, and mortality and the concept of no threshold. In *Advances in Environmental Science and Technology*, Vol. 3, J. W. Pitts and R. L. Metcalf (Eds.), Wiley, New York, pp. 127–156.

Carrie, C., and Schnettler, O. (1939). Prevention of leucopenia after roentgen irradiation. *Strahlenther.*, **66**: 149–154.

Carroll, R. E. (1969). The relationship of cadmium in the air to cardiovascular death rates. *Am. Med. Assoc.*, **198**: 267.

Carson, P. E., Brewer, G. J., and Ickes, C. (1961). Decreased glutathione reductase with susceptibility to hemolysis. *J. Lab. Clin. Med.*, **58**: 804 (abstract).

Carter, L. J. (1976). Toxic substances: Five year struggle for landmark bill may soon be over. *Science*, **194**: 40–42.

Cassidy, J. T., and Nordby, G. T. (1975). Human serum immunoglobulin concentrations: Prevalence of immunoglobulin deficiencies. *J. Allerg. Clin. Immunol.*, **55**: 35–48.

Cecil, H. C., Harris, S. J., Bitman, J., and Fries, G. F. (1973). Polychlorinated biphenyl-induced decrease in liver vitamin A in Japanese quail and rats. *Bull. Environ.*, **1**: 89.

Chadwick, R. W., Cranmer, M. F., and Peoples, A. J. (1971). Metabolic alterations in the squirrel monkey induced by DDT administration and ascorbic acid deficiency. *Toxicol. Appl. Pharmacol.*, **20**: 308–318.

Chadwick, R., Peoples, A., and Cranmer, M. (1973). The effect of protein quality and ascorbic acid deficiency on stimulation of hepatic microsomal enzymes in guinea pigs. *Toxicol. Appl. Pharmacol.*, **24**: 603–611.

Chandra, S. V., and Tandon, S. D. (1973). Enhanced manganese toxicity in iron deficient rats. *Environ. Physiol. Biochem.*, **3**: 230.

Chanlett, E. T. (1973). *Environmental Protection*, McGraw-Hill, New York, p. 236.

Chapman, D. W., and Shaffer, C. F. (1947). Mercurial diuretics. *Arch. Intern. Med.*, **79**: 449–456.

Charles, J. M., and Menzel, D. B. (1975). Ammonium and sulfate ion release of histamine from lung fragments. *Arch. Environ. Health*, **30**: 314–316.

Chia, B. L., Chew, C. H., and Lee, S. (1970). Recurrent respiratory tract infection due to isolated absence of IgA. *Med. J. Malaya*, **24**: 215–217.

Chiemchaisri, Y., and Philips, P. H. (1963). Effect of dietary fluoride upon magnesium calcinosis syndrome. *J. Nutr.*, **81**: 307–311.

Chisolm, J. J. (1962). Aminoaciduria as a manifestation of renal tubular injury in lead intoxication and a comparison with patterns of aminoaciduria seen in other diseases. *J. Pediatr.*, **60**: 1.

Chisolm, M. (1974). The association between webs, iron, and post cricoid carcinoma. *Postgrad. Med. J.*, **50**: 215–219.

Chow, C. K., Dillard, C. J., and Tappel, A. L. (1974). Glutathione peroxidase system and lysozyme in rats exposed to ozone or nitrogen dioxide. *Environ. Res.*, **7**: 311–319.

Chowdhury, P., and Louria, D. B. (1976). Influence of cadmium and other trace metals on human a_1-antitrypsins: An *in vitro* study. *Science*, **191**: 480.

Chu, E. W., and Malmgren, R. A. (1965). An inhibitory effect of vitamin A on the induction of tumors of fore stomach and cervix in the Syrian hamster by carcinogenic polycyclic hydrocarbons. *Cancer Res.*, **25**: 884.

Clarke, E. G. C., and Clarke, M. L. (1967). *Garner's Veterinary Toxicology*, 3rd ed., Williams and Wilkins, Baltimore, pp. 156–158.

Clarkson, T. W., and Kench, J. E. (1956). Urinary excretion of amino acids by men absorbing heavy metals. *Biochem. J.*, **62**: 361.

Clausen, A. (1942). Treatment of x-ray leucopenia with vitamin C. *Acta Radiol.*, **23**: 95–98.

Cleaver, J. E. (1968). Defective repair replication of DNA in xeroderma pigmentosum. *Nature*, **218**: 652–656.

Cleaver, J. E., and Carter, P. M. (1973). Xeroderma pigmentosum: Influence of temperature on DNA repair. *J. Invest. Dermatol.*, **60**: 29–32.

Coburn, R. F. (1970). Endogenous carbon monoxide production. *New Engl. J. Med.*, **282**: 207–209.

Coburn, R. F., Blakemore, W. S., and Forester, R. E. (1963). Endogenous carbon monoxide production in man. *J. Clin. Invest.*, **42**: 1172–1178.

Coffin, D. L., Gardner, D. E., Holzman, R. S., and Wolcocks, F. J. (1968). Influence of ozone on pulmonary cells. *Arch. Environ. Health*, **16**: 633–636.

Cohen, G., and Hochstein, P. (1963). Glutathione peroxidase: The primary agent for the elimination of hydrogen peroxide in erythrocytes. *Biochem.*, **2**: 1420.

Cohen, G., and Hochstein, P. (1964). Generation of hydrogen peroxide in erythrocytes by hemolytic agents. *Biochem.*, **3**: 895.

Cohen, S. I., Perkins, N. M., Ury, H. K., and Goldsmith, J. R. (1971). Carbon monoxide uptake in cigarette smoking. *Arch. Environ. Health*, **22**: 55–60.

Cohn, Z. A., and Wiener, E. (1964). The particulate hydrolases of macrophages: II. Biochemical and morphological response to particle ingestion. *J. Exp. Med.*, **118**: 1009–1023.

Cole, P., and Goldman, M. B. (1975). Occupation. In *Persons at High Risk of Cancer: An Approach to Cancer Etiology and Control*, J. F. Fraumeni, Jr. (Ed.), Academic, New York, pp. 167–183.

Collins-Williams, C., Kokubu, H. L., Lamenza, C., Nizami, R., Chiu, A. W., Lewis-McKinley, C., Comerford, T. A., and Varga, E. A. (1972). Incidence of isolated deficiency of IgA in the serum of Canadian children. *Ann. Allerg.*, **30**: 11–23.

Colucci, A. V., Hammer, D. I., Williams, M. E., Hinners, T. A., Pinkerton, C., Kent, J. L., and Love, G. J. (1973). Pollutant burdens and biological response. *Arch. Environ. Health*, **27**: 151–154.

Committee on Nitrate Accumulation, Agricultural Board, Division of Biology and Agriculture, National Research Council. (1972). Accumulation of Nitrate. National Academy of Sciences, Washington, D.C.

Commoner, B. (1970). Threats to the integrity of the nitrogen cycle: Nitrogen compounds in soil, water, atmosphere and precipitation. In *Global Effects of Environmental Pollution*, S. F. Singer (Ed.), Reidel Pub. Co.

Conray, R. T. W. L., and Mills, J. N. (1970). *Human Circadian Rhythms.* Churchill, London.

Consumer Bulletin (March, 1963). Hard water is best for health.

Cooley, J. C., Peterson, W. L., Engel, C. E., and Jemigan, J. P. (1954). Clinical trait of massive splenic infarction, sicklemia trait and high altitude flying. *J. Am. Med. Assoc.*, **154**: 111.

Cooper, W. C. (1973). Indicators of susceptibility to industrial chemicals. *Jour. Occup. Med.*, **15**(4): 355.

Cooper, W. C., and Tabershaw, I. R. (1966). Biologic effects of nitrogen dioxide in relation to air quality standards. *Arch. Environ. Health,* **12**: 522–530.

Cornblath, M., and Hartman, A. F. (1948). Methemoglobinemia in young infants. *J. Pediatr.*, **33**: 421.

Corrin, B., and King, E. (1970). Pathogenesis of experimental pulmonary alveolar proteinosis. *Thorax*, **25**: 230–236.

Court-Brown, W. M., and Doll, R. (1965). Mortality from cancer and other causes after radiotherapy for ankylosing spondylitis. *Br. Med. J.*, **2**(5474): 1327–1332.

Cramer, K. (1966). Predisposing factors for lead poisoning. *Acta Med. Scand.*, Suppl. 445. **179**: 56.

Creasy, W. A. (1960). The enzymic composition of nuclei isolated from radiosensitive and non-sensitive tissue with special reference to catalase activity. *Biochem. J.*, **77**: 5.

Cripps, D. J., and Curtis, A. C. (1962). Toxic effect of chloroquine on porphyria hepatica. *Arch. Derm.*, **86**: 575.

Curban, G. V. (1951). Multiple cutaneous carcinomatosis in 2 albino brothers. *Rev. Paul. Med.*, **39**: 440.

Curran, G. L., and Azarnoff, D. L. (1958). Inhibition of cholesterol biosynthesis in man. *Arch. Intern. Med.*, **101**: 685.

Curtis, G. W., Algeri, E. J., McBay, A. J., and Ford, R. (1955). The transplacental diffusion of CO: A review and experimental study. *Arch. Pathol.*, **59**: 677–690.

Dahl, L. K. (1972). Salt and hypertension. *Am. J. Clin. Nutr.*, **25**: 231.

Dahl, L. K., Leitl, G., and Heine, M. (1972). Influence of dietary potassium and sodium/potassium molar ratios of the development of salt hypertension. *J. Exp. Med.*, **136**: 318–330.

Dahl, L. K., and Love, R. A. (1954). Evidence for relationship between sodium (chloride) intake and human essential hypertension. *Arch. Intern. Med.*, **94**: 525–531.

Dahl, L. K., and Love, R. A. (1957). Etiological role of sodium chloride intake in essential hypertension in humans. *J. Am. Med. Assoc.*, **164**: 397–400.

Dameshek, W., and Singer, K. (1941). Familial non-hemolytic jaundice: Constitutional hepatic dysfunction with indirect Van den Bergh reaction. *Arch. Intern. Med.*, **67**: 259.

Das, M. L., Orrenius, S., and Ernster, L. (1968). On the fatty acid and hydrocarbon hydroxylation in rat liver microsomes. *Eur. J. Biochem.*, **4**: 519.

Dastur, D. K., Manghani, D. K., and Roghavendian, K. V. (1971). Distribution and fate of MN⁵⁴ in the monkey: Studies of different parts of the central nervous system and other organs. *J. Clin. Invest.*, **50**: 9.

Davies, D. P. (1973). Plasma osmolality and feeding practices of healthy infants in first three months of life. *Br. Med. J.*, **2**: 340–342.

Davies, R. E. (1967). Effect of vitamin A on 7,12-dimethylbenz (a) anthracene–induced papillomas in rhino mouse skin. *Cancer Res.*, **27**: 237.

Davies, R. O., Marton, A. V., and Kalow, W. (1960). The action of normal and atypical cholinesterase of human serum upon a series of esters of choline. *Can. J. Biochem. Physiol.*, **38**: 545.

Dawson, J. P., Thayer, W. W., and Desforges, J. F. (1958). Acute hemolytic anemia in the newborn infant due to naphthalene poisoning: Report on two cases with investigation into the mechanism of the disease. *Blood*, **13**: 1113.

Dean, G., and Barnes, H. D. (1959). Porphyria in Sweden and South Africa. *S. Afr. Med. J.*, **33**: 274.

Dean, G. (1963). *The Porphyrias: A Story of Inheritance and Environment*, Pitman, London.

deLeeuw, W. K. M., Lowenstein, L., and Hsieh, Y. S. (1966). Iron deficiency and hydremia in normal pregnancy. *Medicine*, **45**: 291.

Deringer, M. K., Dunn, T. B., and Heston, W. E. (1953). Results of exposure of strain C3H mice to chloroform. *Proc. Soc. Exp. Biol. Med.*, **83**: 474–479.

Diamond, E. L., Schmerler, H., and Lilienfield, A. (1973). The relationship of intra-uterine radiation to subsequent mortality and development of leukemia in children. *Am. J. Epidemiol.*, **97**(5): 238–313.

Dianzani, M. U. (1963). Lysosome changes in liver injury. In *Ciba Symposium on Lysosomes*, A. V. S. de Reuck and M. P. Camerson (Eds.), Little, Brown, Boston, pp. 335–352.

DiBenedetto, R. J. (1967). Chronic hypervitaminosis A in an adult. *J. Am. Med. Assoc.*, **201**: 700.

Dietrich, G., and Buchner, M. (1960). Contribution to the vitamin C metabolism of smokers. *Dtsch. Gesundeheitwes.*, **15**: 2494–2495.

Dinning, J. S. (1953). Some effects of vitamin E on amino acid metabolism. *Fed. Proc.*, **12**: 412.

Dolgova, Z. Ya. (1962). Ascorbic acid exchange during the action of x-rays on the organism. *Med. Radiol.*, **7**: 67–70.

Done, A. K. (1964). Developmental pharmacology. *Clin. Pharmacol. Ther.*, **5**: 432–479.

Douglas, S. D., Goldberg, L. S., Fudenberg, H. H., and Goldberg, S. B. (1970). Agammaglobulinaemia and co-existent pernicious anemia. *Clin. Exp. Med.*, **127**: 181–187.

Drill, V. A. (1952). Hepatotoxic agents: Mechanism of action and dietary interrelationship. *Pharmacol. Rev.*, **4**: 1–42.

Dubois, E. L. (1974). *Lupus Erythematosus*, 2nd ed., University of Southern California Press, Los Angeles, p. 798.

Dubreuilh, W. (1896). Des Hyperkeratoses Circonscriptes. *Ann. Dermatol. Syph.*, Ser. 3. **7**: 1158–1204.

Dungal, N., and Sigurjonsson, J. (1967). Gastric cancer and diet: A pilot study on dietary habits in two districts differing markedly in respect of mortality from gastric cancer. *Br. J. Cancer*, **21**: 270–276.

Durand, C. H., et al. (1962). Latent hypovitaminosis and tobacco. *Concourse Med.*, **84**: 4801–4806.

Durham, W. F. (1967). The interaction of pesticides with other factors. *Residue Rev.*, **8**: 21.

Dutton, G. J. (1961). The mechanism of glucuronide formation. *Biochem. Pharmacol.*, **6**: 65.

Dutton, G. J. (1962). Glucuronide conjugation. *Proc. 1st Int. Pharmacol. Meet.*, **6**: 39.

Dutton, G. J. (1966). Free and combined. In *Glucuronic Acid*, C. J. Dutton (Ed.), Academic, New York, pp. 222, 230.

Eales, L. (1963). Porphyria as seen in Cape Town: A survey of 250 patients and some recent studies. *S. Afr. J. Lab. Clin. Med.*, **9**: 151.

Eales, L., and Linder, G. C. (1962). Porphyria: The acute attack. *S. Afr. Med. J.*, **36**: 284.

Easton, R. E., and Murphy, S. D. (1967). Experimental ozone exposure and histamine. *Arch. Environ. Health*, **15**: 100.

Ehrlich, R. (1966). Effect of nitrogen dioxide on respiratory infection. *Bacteriol. Rev.*, **30**: 604–614.

Eisen, A. Z., Block, K. J., and Sakai, T. (1970). Inhibition of human skin collagenase by human serum. *J. Lab. Clin. Med.*, **75**: 258.

Eisen, H. (1974). *Immunology: An Introduction to Molecular and Cellular Principles of the Immune Response*, Harper & Row, New York, pp. 473–474, 504.

Elliot, G. B., and Alexander, E. A. (1961). Sodium from drinking water as an unsuspected cause of cardiac decompensation. *Circulation*, **23**: 562.

Emmelot, P., and Benedetti, E. L. (1960). Changes in the fine structure of rat liver brought about by dimethylnitrosamine. *J. Biophys. Biochem. Cytol.*, **7**: 393–396.

Environmental Health Resource Center. (1973). Health effects and recommended standard for atmospheric asbestos. Illinois Institute for Environmental Quality. IIEQ doc. no. 73–2.

Environmental Health Resource Center. (1974). Advisory report on health effects of nitrates in water. Illinois Institute for Environmental Quality. IIEQ doc. no. 74–5.

Environmental Health Resource Center. (1975). Health effects and recommended alert and warning system for ozone. Illinois Institute for Environmental Quality. IIEQ doc. no. 75–17.

Epstein, J. H. (1966). Ultraviolet light carcinogenesis. In *Advances in Biology of Skin*, Vol. VII, *Carcinogenesis*, W. Montagna and R. L. Dobson (Eds.), Appleton-Century-Crofts, New York.

Epstein, W. L., Fukuyama, K., and Epstein, J. H. (1971). UV light, DNA repair and skin carcinogenesis in man. *Fed. Proc.*, **30**: 1766–1771.

Eschenbrenner, A. B., and Miller, E. (1944). Induction of hepatomas in mice by repeated oral administration of chloroform, with observations on sex differences. *J. Nat. Cancer Inst.*, **5**: 251–255.

Evans, F. T., Gray, P. W. S., Lehmann, H., and Silk, E. (1952). Sensitivity to succinylcholine in relation to serum cholinesterase. *Lancet*, **1**: 1229.

Evans, R. D. (1967). The radium standard for boneseekers: Evaluation of the data on radium patients and dial painters. *Health Phys.*, **13**: 267.

Evans, T. C. (1947). Effects of hydrogen peroxide produced in the medium by radiation on spermatozoa of *Arbacia punctulata. Biol. Bull.*, **92**: 99.

Fagerhol, M. K. (1973). Recent findings and ideas concerning the Pi polymorphism and diseases associated with alpha₁ -antitrypsin deficiency. In *Fundamental Problems of Cystic Fibrosis and Related Diseases*, Intercontinental, New York, p. 402.

Fagerhol, M. K., and Braend, M. (1965). Serum prealbumins: Polymorphism in man. *Science*, **149**: 986.

Fagerhol, M. K., and Laurell, C. B. (1970). The Pi system: Inherited variants of serum alpha₁-antitrypsin. In *Progress in Medical Genetics*, Vol. 7, A. G. Steinberg and A. G. Bearn (Eds.), Grune and Stratton, New York, p. 96.

Fahey, J. L. (1971). Cancer in the immunosuppressed patient. *Ann. Intern. Med.*, **75**: 310–312.

Fairbairn, A. S., and Reid, D. D. (1958). Air pollution and other local factors in respiratory disease. *Br. J. Prev. Soc. Med.*, **12**: 94.

Fairbanks, V. F., and Beutler, E. (1962). A simple method for detection of erythrocyte glucose-6-phosphate dehydrogenase (G-6-PD Spot Test). *Blood*, **20**: 591.

Fears, T. R., Scotto, J., and Schniederman, M. A. (1976). Skin cancer, melanoma, and sunlight. *Am. J. Pub. Health*, **66**(5): 461–464.

Feinstein, R. N., Faulhaber, J. T., and Howard, J. B. (1968). Sensitivity of acatalesemic mice to acute and chronic irradiation and related conditions. *Radiat. Res.*, **35**: 341–349.

Felger, G. (1952). Relationship between reduced glutathione content and spontaneous hemolysis in shed blood. *Nature*, **170**: 624.

Felton, G., and Patterson, M. G. (1971). Shift rotation is against nature. *Am. J. Nurs.*, **71**: 4, 760.

Fennelly, J. J., Fitzgerald, O., and Hingerty, D. J. (1960). Observations on porphyria with special reference to Ireland. *Ir. J. Med. Sci.*, **411**: 130.

Feuer, G., and Liscio, A. (1970). Effect of drugs on hepatic drug metabolism in the fetus and newborn. *Int. J. Clin. Pharmacol.*, **31**: 30–33.

Filippov, Y. M. (1964). The biosynthesis of ascorbic acid in the organs of white rats treated with dipterex (0-0, 0-dimethyl-2, 2,2-trichloro-1-hydroxyethylphosphate, cholinesterase inhibitor). *Gig. Sanit.*, **5**: 30–34.

Finklea, J. F., Shy, C. M., Moran, S. B., Nelson, W. C., Larsen, R. I., and Akland, G. G. (1974). The role of environmental health assessment in the control of air pollution. Paper presented to the 67th Meeting of the American Institute of Chemical Engineers, Houston.

Firket, J., Secretary. (1931). Sur les causes des accidents servenus dans la Valler de la Neuse, Lors des Bronillards de December, 1930. Resultats de l'expertise. *Bull. Acad. R. Med. Belg.*, **11**: 683–741.

Fitzpatrick, T. B., and Quevedo, W. C. (1972). Albinism. In *The Metabolic Basis of Inherited Disease*, 3rd ed., J. B. Stanbury, J. B. Wyngaarden, and D. S. Fredrickson (Eds.), McGraw-Hill, New York, pp. 326–337.

Fletcher, B. L., and Tappel, A. L. (1972). Protective effects of dietary a-tocopherol in rats exposed to toxic levels of ozone and nitrogen dioxide. *Environ. Res.*, **6**: 165–175.

Flick, D. F., Kraybill, H. F., and Dimitroff, S. M. (1971). Toxic effects of cadmium: A review. *Environ. Res.*, **4**: 71–85.

Forbes, G. B., and Reina, J. C. (1972). Effect of age on gastrointestinal absorption (Fe, Sr, Pb) in the rat. *J. Nutr.*, **102**: 647.

Forssman, S., and Frykholm, K. U. (1947). Benzene poisoning: II. Examination of the workers exposed to benzene with reference to the presence of estersulfate, muconic acid, urochrome A and polyphenol in the urine together with vitamin C deficiency. Prophylactic measures. *Acta Med. Scand.*, **128**(3): 256–280.

Fosburgh, L. (1974). Bad air's effect on young assayed. *New York Times*, Dec. 12.

Fourman, P., and Morgan, D. B. (1962). *Proc. Nutr. Soc.*, **21**: 34.

Fouts, J. R., and Adamson, R. H. (1959). Drug metabolism in the newborn rabbit. *Science*, **129**: 897–898.

Fox, M. R., and Fry, B. E. (1970). Cadmium toxicity decreased by dietary ascorbic acid supplements. *Science*, **169**: 989–991.

Freeman, G., Stephens, R. J., Crane, S. C., et al. (1968). Lesion of the lungs in rats continuously exposed to two parts per million of nitrogen dioxide. *Arch. Environ. Health*, **17**: 181–192.

Freeman, G., and Haydon, G. B. (1964). Emphysema after low-level exposure to NO_2. *Arch. Environ. Health*, **8**: 125–128.

Freeman, G., Hudson, A., Carnes, R., et al. (1970). Salicylanilide photosensitivity. *J. Invest. Dermatol.*, **54**: 145–149.

Freis, E. D. (1960). Hemodynamics of hypertension. *Physiol. Rev.*, **40**: 27.

Freis, E. D. (1976). Salt, volume and the prevention of hypertension. *Circulation*, **58**(4): 589–595.

Friberg, L. T., Piscator, M., and Nordberg, G. F. (1972). *Cadmium in the Environment*. Chemical Rubber Co., Cleveland, Ohio.

Friedman, J. J. (1971). Functional properties of blood. In *Physiology*, E. E. Sulkurt (Ed.), Little, Brown, Boston, p. 223.

Friend, B. (Nov., 1970). *National Food Situation*. U.S. Government Printing Office.

Friend, M., and Trainer, D. O. (1970). Polychlorinated biphenyl: Interaction with duck hepatic virus. *Science*, **170**: 1314.

Froggatt, P. (1957). Albinism: A statistical, genetical and clinical appraisal based upon a complete ascertainment of the condition in Northern Ireland. Thesis, Trinity College, Dublin, Ireland.

Frommel, D., Moullec, J., Lambin, P., and Fine, J. M. (1973). Selective serum IgA deficiency: Frequency among 15,200 French blood donors. *Vox. Sang.*, **25**: 513–518.

Furstenberg, A. C., Richard, G., and Lathrop, F. D. (1941). Ménière's disease: Addenda to medical therapy. *Arch. Otolaryngol.*, **34**: 1083–1092.

Galdston, M., Gottwelt, C., and Davies, A. L. (1972). Leucocyte lysosomal enzymes and a_1-antitrypsin: Genetic determinants in chronic obstructive pulmonary disease. *Fed. Proc.* (abstract).

Ganote, C. E., and Rosenthal, A. L. (1968). Characteristic lesions of methyl-azoxymethanol-induced liver damage. *Lab. Invest.*, **19**: 382–389.

Ganther, H. E., Gordie, C., Sunde, M. L., Kopecky, M. J., Wagner, P., Sang Hwan Oh, and Hoekstra, W. G. (1972). Selenium: Relation to decreased toxicity of methylmercury added to diets containing tuna. *Science*, **175**: 1122.

Gardner, E. E., Rogler, J. C., and Parker, H. E. (1961). Interrelationships between magnesium and fluoride in chicks. *J. Nutr.*, **75**: 270.

Gardner, E. J. (1972). *Principles of Genetics.* Wiley, New York, p. 390.

Garfinkel, D. (1958). Studies on pig liver microsomes: I. Enzymatic and pigment composition of different microsomal fractions. *Arch. Biochem. Biophys.*, **77**: 493–509.

Garrington, G. E., Scofield, H. H., Cornyn, J., and Lacy, G. R. (1967). Intraoral malignant melanoma in a human albino. *Oral Surg.*, **24**: 224.

Garther, L. M., and Arias, I. M. (1966). Studies of prolonged neonatal jaundice in the breast-fed infant. *J. Pediatr.*, **68**(1): 54.

Gelzayd, E. A., McCleery, J. L., Melnyk, C. S., and Kraft, S. C. (1971). Intestinal malabsorption and immunoglobulin deficiency. *Arch. Intern. Med.*, **127**: 141–147.

Genazzani, E., and Miele, E. (1959). Ionizing radiations and lysozyme: II. *Boll. Soc. Ital. Biol. Sper.*, **35**: 1798–1801.

Gentz, J., Jagenburg, R., and Zetterstrom, R. (1965). Tyrosinemia: An inborn error of tyrosine metabolism with cirrhosis of the liver and multiple renal tubular defects (Le Toni-Debré-Fanconi syndrome). *J. Pediatr.*, **66**: 670.

Ghoshal, A. K., Porta, E. A., Hartroft, W. S. (1969). The role of lipoperoxidation in the pathogenesis of fatty livers induced by phosphorous poisoning in rats. *Am. J. Pathol.*, **54**: 275–291.

Giblett, E. R. (1969). *Genetic Markers in Human Blood.* Blackwell, Oxford, p. 192.

Gillette, J. R. (1967). Individually different responses to drugs according to age, sex and functional or pathological states. *Drug Responses in Man*, Churchill, London, p. 24.

Ginsberg, A., and Mullinax, F. (1970). Pernicious anemia and monoclonial gammopathy in a patient with IgA deficiency. *Am. J. Med.*, **48**: 787–791.

Glaser, C. B., Karic, L., Huffaker, T., and Fallat, R. J. (1977). Influence of cadmium on human alpha-1-antitrypsin: a reexamination. *Science*, **196**: 556–557.

Goedde, H. W., Altland, K. J., and Schloot, W. (1968). Therapy of prolonged apnea after suxamethonium with purified pseudocholinesterase: New data on kinetics of the hydrolysis of succinyldicholine and succinylmonocholine and further data on *N*-acetyltransferase-polymorphism. *Ann. N.Y. Acad. Sci.*, **151**: 742.

Golberg, L. (Ed.). (1974). The toxicity of polychlorinated polycyclic compounds and related chemicals. *Crit. Rev. Toxicol.*, **2**(4): 445.

Goldberg, A. (1968). Lead poisoning as a disorder of heme synthesis. *Semin. Hematol.*, **5**: 424.

Goldsmith, M. F., and Goldsmith, J. (1966). Epidemiological aspects of magnesium and calcium metabolism. *Arch. Environ. Health*, **12**: 607.

Goldstein, A. L. (July, 1975). The role of thymosin in the immune response. Presented at the Frederick Cancer Research Center, Frederick, Md.

Goldstein, A. L., Hooper, J. A., Schuolf, R. S., Cohen, G. H., Thurman, G. B., McDaniel, M. D., White, A., and Dardenne, M. (1974). Thymosin and immunopathology of aging. *Fed. Proc.*, **33**(9): 2053.

Goldstein, A. L., Thurman, G. B., Cohen, G. A., and Hooper, J. A. (1975). *Biological Activity of Thymic Hormones*, Kooyker Scientific Publications, Rotterdam, The Netherlands.

Goldstein, A. L., Wara, D. W., Amman, A. J., Sakai, H., Harris, W. S., Thurman, G. B., Hooper, J. A., Cohen, G. H., Goldman, H. S., Constanzi, J. J., and McDaniel, M. D. (1975). First clinical trial with thymosin: Reconstruction of T-cells in patients with cellular immuno deficiency diseases. *Transplant. Proc.*, **7**(1): 681.

Goldstein, A. L., and White, A. (1971). Thymosin and other thymic hormones: Their nature and roles in the thymic dependency of immunological phenomena. In *Contemporary Topics in Immunobiology*, A. J. S. Davis and R. L. Carter (Eds.), Plenum, New York, pp. 339–350.

Goldstein, B. D. (1973). Hydrogen peroxide in erythrocytes. *Arch. Environ. Health*, **26**: 279.

Goldstein, B. D., and Balchum, O. J. (1967). Effect of ozone on lipid peroxidation in the red blood cell. *Proc. Soc. Exp. Biol. Med.*, **126**: 356.

Goldstein, B. D., Buckley, R. D., Cardenas, R., and Balchum, O. J. (1970). Ozone and vitamin E. *Science*, **169**: 605.

Goldstein, E., and Green, G. M. (1966). The effects of acute renal failure on the bacterial clearance mechanisms of the lung. *J. Lab. Clin. Med.*, **68** (4): 531.

Gordon, H. H., Nitowsky, H. M., Tildon, B. S., and Levin, S. (1958). Studies of tocopherol deficiencies in infants and children: V. An interim summary. *Pediatrics*, **21**: 673.

Gontzea, J. (1963). The vitamin C requirements of lead workers. *Int. Z. Augenwardte Physiol. Einschl. Arb. Physiol.*, **20**: 20–33.

Gould, S., Allison, H. M., and Bellew, L. N. (1961). Acute porphyria complicated by pregnancy: Report of a case. *Obstet. Gynecol.*, **17**: 109.

Goyanna, C. (1955). Tobacco and vitamin C. *Bras. Med.*, **69**: 173–177.

Graham, S. (Oct. 21, 1972). Editorial, *Br. Med. J.*, **4**: xx.

Graham, S., Schotz, W., and Martino, P. (1972). Alimentary factors in the epidemiology of gastric cancer. *Cancer*, **30**: 927–938.

Granick, S. (1966). The induction *in vitro* of the synthesis of δ-aminolevulinic acid synthetase in chemical porphyria: A response to certain drugs, sex hormones and foreign chemicals. *J. Biol. Chem.*, **241**: 1359.

Greenberg, M. A., Nelson, K. E., and Carnow, B. W. (1973). A study of the relationship between sudden infant death syndrome and environmental factors. *Am. J. Epidemiol.*, **98**(6): 412–422.

Greenberg, M. S., Kass, E. H., and Castle, W. B. (1957). Studies on the destruction of red blood cells: XII. Factors influencing the role of hemoglobin in the pathologic physiology of sickle cell anemia and related disorders. *J. Clin. Invest.*, **36**: 833.

Greenblatt, A. (1974). *Heredity and You*, Coward, McCann and Geoghegan, New York, pp. 155–166.

Greenblatt, J. J. (1957). Use of massive doses of vitamin E in humans and rabbits to reduce blood lipids. *Circulation*, **16**: 508.

Gruener, N., and Shuval, H. I. (1970). Health aspects of nitrates in drinking water. In *Developments in Water Quality*, H. I. Shuval (Ed.), Humphrey, Ann Arbor, Mich., pp. 89–106.

Gunn, S. A., Gould, T. C., and Anderson, W. A. D. (1968). Specificity in protection against lethality by cadmium. *Proc. Soc. Exp. Biol. Med.*, **128**: 591.

Guy, W. S., and Taves, D. R. (1973). *Int. Assoc. Dent. Res. Abstr.*, **718**.

Guyton, A. C., Coleman, T. G., Cawley, A. W., Manning, R. D., Jr., Norman, R. A., and Ferguson, J. D. (1974). A systems analysis approach to understanding long-range arterial blood pressure control and hypertension. *Circ. Res.*, **35**: 159.

Haddy, F. J., and Overbeck, H. W. (1976). The role of humoral agents in volume expanded hypertension. *Life Sci.*, **19**: 935–948.

Haeger-Aronsen, B. (1960). Studies on urinary excretion of δ-aminolaevulic acid and other haem precursors in lead workers and lead-intoxicated rabbits. *Scand. J. Clin. Invest.*, Suppl. 47. **12** (cited in Marver and Schmid, 1972).

Haenszel, W. (1967). Epidemiology of gastric cancer. In *Neoplasms of the Stomach*, C. McNeer and G. T. Pach (Eds.), Lippincott, Philadelphia, pp. 3–28.

Haenszel, W., Kurihara, M., Segi, M., et al. (1972). Stomach cancer among Japanese in Hawaii. *J. Natl. Cancer Inst.*, **49**: 969–988.

Halberg, F. (1960). The 24-hour scale: A time dimension of adaptive functional organization. *Perspect. Bio. Med.*, **3**: 4, 491.

Halberg, F., Bittner, J. J., Gully, R. J., Albrecht, P. G., and Brackney, E. L. (1955). 24-hour periodicity and audiogenic convulsions in I mice of various ages. *Proc. Soc. Exp. Biol. Med.*, **88**: 169.

Halberg, F., Jacobsen, E., Wadsworth, G., and Bittner, J. J. (1958). Audiogenic abnormality spectra, twenty-four hour periodicity and lighting. *Science*, **128**: 657.

Halberg, F., Johnson, E. A., Brown, B. W., and Bittner, J. J. (1960). Susceptibility rhythm to *E. Coli* endotoxin and bio-assay. *Proc. Soc. Exp. Soc. Med.*, **103**: 142.

Hall, G. E., and Lucas, C. C. (1937). Choline-esterase activity of normal and pathological serum. *J. Pharmacol. Exp. Ther.*, **59**: 34.

Hall, J. C., Wood, C. H., Stoeckle, J. D., and Tepper, L. P. (1959). Case data from the beryllium registry. *AMA Mich. Ind.*, **19**: 100.

Hallet, W. Y., Knudson, A. G., Jr., and Massey, F. J., Jr. (1965). Absence of detrimental effect of the carrier state for the cystic fibrosis gene. *Am. Rev. Respir. Dis.*, **92**: 714.

Hammond, E. C., and Selikoff, I. J. (1973). Relations of cigarette smoking to risk of death of asbestos-associated disease among insulation workers in the United States. In *Biological Effects of Asbestos*, International Agency of Research on Cancer, Lyon, pp. 312–317.

Hammond, R. S., and Welcker, M. L. (1948). Porphobilinogen tests on a thousand miscellaneous patients in a search for false positive reactions. *J. Lab. Clin. Med.*, **33**: 1254.

Hanhijärvi, H. (1974). Comparison of free ionized fluoride concentrations of plasma and renal clearance in patients of artificially fluoridated and non-fluoridated drinking water areas. *Proc. Finn. Dent. Soc.*, Suppl. III. **70**: 12.

Hans, E., and Halberg, F. (1959). 24-hour rhythm in susceptibility of C mice to a toxic dose of ethanol. *J. Appl. Physiol.*, **14**: 878.

Hans, E., Hanton, E. M., and Halberg, F. (1959). 24-hour susceptibility rhythm to ethanol in fully fed, starved, and thirsted mice and the lighting regimen. *Physiologist*, **2**: 54.

Han-Wen, H., Lu, S. H., and Chang, M. C. (1959). Treatment of lead poisoning: II. Experiments on the effect of vitamin C and rutin. *Chin. J. Intern. Med.*, **7**: 19–20.

Hardy, H. L., and Stoeckle, J. D. (1959). Beryllium disease. *J. Chronic Dis.*, **9**: 152.

Hardy, M. A., Zisblatt, M., Levine, N., Goldstein, A. L., Lilly, F., and White, A. (1971). Reversal by thymosin of increased susceptibility of immunosuppressed mice to Maloney sarcoma virus. *Transplant. Proc.*, **3**: 926.

Harman, D. (1963). Role of serum copper in coronary atherosclerosis. *Circulation*, **28**: 658.

Harman, D. (1968). Atherogenesis in minipigs: Effect of dietary fat unsaturation and of copper. *Circulation*, Suppl. VI. **38**: 8.

Harmeson, R. H., Sollo, F. W., Jr., and Larson, T. E. (1971). The nitrate situation in Illinois. *J. Am. Water Works Assoc.*, **63**: 303.

Harris, H., and Whittaker, M. (1961). Differential inhibition of human serum cholinesterase with fluoride: Recognition of two phenotypes. *Nature*, **191**: 496.

Harris, H., and Robson, E. B. (1963). Screening tests for the "atypical" and "intermediate" serum cholinesterase variant. *Lancet*, **ii**: 218.

Harris, P. L., Hardenbrook, E. G., Dean, F. P., Cusack, E. R., and Jensen, J. L. (1961). Blood tocopherol values in normal adults and incidence of vitamin E deficiency. *Proc. Soc. Exp. Biol. Med* **107**: 381.

Harrison, H. E., and Harrison, H. C. (1954). Experimental production of renal glycosuria, phosphaturia and aminoaciduria by injection of maleic acid. *Science*, **120**: 606.

Hathaway, J. A., and Terrill, R. E. (1962). Metabolic effect of chronic ozone exposure on rats. *Am. Ind. Hyg. Assoc. J.*, **23**: 392–395.

Hermans, P. E. (1967). Nodular lymphoid hyperplasia of the small intestine and hypogammaglobulinemia: Theoretical and practical considerations. *Fed. Proc.*, **26**: 1606–1611.

Herndon, J. H., Jr., and Freeman, R. G. (1976). Human disease associated with exposure to light. *Ann. Rev. Med.*, **27**: 77–87.

Hewitt, H. B. (1956). Renal necrosis in mice after accidental exposure to chloroform. *Br. J. Exp. Pathol.*, **37**: 32–39.

Hickey, R. J., Boyce, D. E., Harner, E. S., et al. (1970). Ecological statistical studies concerning environmental pollution and chronic disease. *IEEE Trans. Geosci. Elect.*, **GE-8**: 186–202.

Hickey, R. J., Clelland, R. C., Bower, E. J., and Boyce, D. E. (1976). Health effects of atmospheric sulfur dioxide and dietary sulfite. *Arch. Environ. Health*, March/April, pp. 108–110.

Hill, R. N., Clemens, T. L., Liu, D. K., Vesell, E. S., and Johnson, W. D. (1975). Genetic control of chloroform toxicity in mice. *Science*, **190**: 159–161.

Hochman, A., and Block-Frankenthal, I. (1953). The effect of low and high x-ray dosage on the ascorbic acid content of the suprarenal. *Br. J. Radiol.*, **26**: 599–600.

Hoekstra, W. G. (1974). Biochemical role of selenium in trace element metabolism in animals. In *Trace Element Metabolism in Animals*, Vol. 2, W. G. Hoekstra, J. W. Suttie, H. Ganther, and W. Mertz (Eds.), University Park Press, Baltimore, p. 61.

Holden, H. (1965). Cadmium fume. *Ann. Occup. Hyg.*, **8**: 51–54.

Holmberg, R. E., and Ferm, V. H. (1969). Interrelationships of selenium, cadmium and arsenic in mammalian teratogenesis. *Arch. Environ. Health*, **18**: 873.

Holmes, H. N., Campbell, K., and Amberg, E. J. (1939). Effect of vitamin C on lead poisoning. *J. Lab. Clin. Med.*, **24**: 1119–1127.

Holtzman, R. B., and Ilcewicz, F. H. (1966). Lead-210 and polonium-210 in tissues of cigarette smokers. *Science*, **153**: 1259–1260.

Horwitt, M. K. (1960). Vitamin E and lipids metabolism in man. *Am. J. Clin. Nutr.*, **8**: 451.

Huling, E. E., and Hollocher, T. C. (1972). Groundwater contamination by road salt: Steady-state concentrations in east central Massachusetts. *Science*, **176**: 288–290.

Hult, H. (1950). "Cholémie simple familiale" (Gilbert) and posthepatic states without fibrosis of liver. *Acta Med. Scand.*, Suppl. 244. **138**: 1.

Hunt, V. R. (1975). Occupational health problems of pregnant women. U.S. Dept. HEW. Order no. SA-5304-75.

Hutton, J. J. (1971). Genetic regulation of glucose-6-phosphate dehydrogenase activity in the inbred mouse. *Biochem. Genet.*, **5**: 315.

Hutzinger, O., Nash, D. M., Safe, S., Defreitas, A. S. W., Nortstrom, R. J., Wildish, D. V., and Zitko, V. (1972). Polychlorinated biphenyls: Metabolic behavior of pure isomers in pigeons, rats and brook trout. *Science*, **178**: 312.

Illinois Pollution Control Board (May, 1976). Ozone episode regulations.

Jacob, H. S., Ingbar, S. H., and Jandl, J. H. (1965). Oxidative hemolysis and erythrocyte metabolism in hereditary acatalasemia. *J. Clin. Invest.*, **44**: 1187.

Jacob, H. S., and Jandl, J. H. (1966). A simple visual screening test for G-6-PD deficiency employing ascorbate and cyanide. *New Engl. J. Med.*, **274**: 1162.

Jacobs, R. M., Fox, M. E., and Aldridge, M. H. (1969). Changes in plasma proteins associated with the anemia produced by dietary cadmium in Japanese quail. *J. Nutr.*, **99**: 119.

Jacobsen, L., Krag Anderson, E., Thorborg, J. V. (1964). Accidental chloroform nephrosis in mice. *Acta Pathol. Microbiol. Scand.*, **61**(4): 503–513.

Jacobsson, K. (1955). Studies on the trypsin and plasmin inhibitors in human blood serum. *Scand. J. Clin. Lab. Invest.*, Suppl. 14. **7**: 55–102.

Jaffe, E. R. (1966). Hereditary methemoglobinemias associated with abnormalities in the metabolism of erythrocytes. *Am. J. Med.*, **41**: 786.

Jaksch, J. A. (1975). Some economic damages to human health resulting from the catalytic converter. Presented at the 68th Annual Meeting of the Air Pollution Control Association, Boston, pp. 1–23.

James, K. M., Collins, M. L., and Fudenberg, H. H. (1966). A semi-quantitative procedure for estimating serum antitrypsin levels. *J. Lab. Clin. Med.*, **67**: 528.

Jandl, J. H., and Cooper, R. A. (1972). Hereditary spherocytosis. In *The Metabolic Basis of Inherited Disease*, 3rd ed., J. B. Stanbury, J. B. Wyngaarden and D. S. Fredrickson (Eds.), McGraw-Hill, New York, p. 1323.

Jankelson, O. M., Vitale, J. J., and Hegsted, D. M. (1959). Serum magnesium, cholesterol, and lipoproteins in patients with atherosclerosis and alcoholism. *Am. J. Clin. Nutr.*, **7**: 23.

Janoff, A. (1972). Elastase-like proteases of human granulocytes and alveolar macrophages. In *Pulmonary Emphysema and Proteolysis*, Charles Mittman (Ed.), Academic, New York, p. 562.

Jeffries, H. E., Sickles, J. E., and Ripperton, L. A. (1976). Ozone transport phenomena: Observed and simulated. Presented at the 69th Annual Meeting of the Air Pollution Control Association, Portland, Ore., June 27–July 1.

Jensen, W. W. (1962). Heredity and chemically induced anemia. *Arch. Environ. Health*, **5**: 43.

Jocelyn, P. C. (1958). The importance of thiol compounds in the causation of disease. *Clin. Chem. Acta*, **3**: 401.

Johnson, H. S. (1973). Report LBL-2217. Lawrence Berkeley Laboratory, Berkeley, Calif.

Jonderko, G. (1961). Diagnostic value of the determination of the blood glutathione level in chronic lead poisoning in human subjects. *Pol. Arch. Med. Wewn.*, **31**: 647.

Jondorf, W. R., Maickel, R. T., and Brodie, B. B. (1958). Inability of newborn mice and guinea pigs to metabolize drugs. *Biochem. Pharmacol.*, **1**: 352–354.

Jones, H. B., and Grendon, A. (1975). Environmental factors in the origin of cancer and estimation of the possible hazard to man. *Food Cosmet. Toxicol.*, **13**: 251–268.

Jones, L. M. (1965). *Veterinary Pharmacology and Therapeutics*, 3rd ed., Iowa State University Press, Ames, pp. 442–447.

Jones, S. R., Binder, R. A., and Donowho, E. M. (1970). Sudden death in sickle-cell trait. *New Engl. J. Med.*, **282**: 232.

Jonxis, J. H. P., and Huisman, T. H. U. (1954). Amino aciduria and ascorbic acid deficiency. *Pediatr.*, **14**, 238.

Jonxis, J. H. P., and Wadman, S. K. (1950). De vitscheiding van aminozuren in de urine bij een patient met scorbuut. *Maandschr. Kindergeneeskd.*, **81**: 251.

Kagan, E., Beissner, I., Gluckman, J., Kuba, P., Cochrane, J. C., Solomon, A., and Webster, I. (1976). Cancer related to asbestos exposure: Immunological studies of patients at risk. *Third International Symposium on Detection and Prevention of Cancer*, Apr. 26–May 1. New York. Symposium 36, no. 511 (abstract).

Kalnins, V. (1953). The effect of x-ray irradiation on the mandibles of guinea pigs treated with large and small doses of ascorbic acid. *J. Dent. Res.*, **32**: 177–188.

Kalow, W., and Davies, R. O. (1958). The activity of various esterase inhibitors towards atypical human serum cholinesterase. *Biochem. Pharmacol.*, **1**: 183.

Kalow, W., and Genest, K. (1957). A method for the detection of typical forms of human cholinesterase: Determination of dibucaine numbers. *Can. J. Biochem. Physiol.*, **35**: 339.

Kalow, W., and Gunn, D. R. (1958). Some statistical data on atypical cholinesterase of human serum. *Ann. Hum. Genet.*, **23**: 239.

Kamm, L., McKeown, G. C., and Smith, D. M. (1965). New colorimetric method for determination of the nitrate and nitrite content of baby foods. *A.O.A.C.*, **48**: 892 (cited in Commoner, 1970).

Kanabrocki, E. L., Case, L. F., Fields, T., et al. (1965). Manganese and copper levels in human urine. *J. Nucl. Med.*, **6**: 780–791.

Kanabrocki, E. L., Fields, T., Decker, C. F., Case, L. F., Miller, E. B., Kaplan, E., and Oester, Y. T. (1964). Neutron activation studies of biological fluids: Manganese and copper. *INT. J. Appl. Radiat. Isot.*, **15**(4): 175–190.

Kar, A. B., Das, R. P., and Karkun, J. N. (1959). Ovarian changes in prepubertal rats after treatment with cadmium chloride. *Acta Biol. Med. Ger.*, **3**: 372.

Kasbekar, D. K., Lavate, W. V., Rege, D. V., and Screenivasan, A. (1959). A study of vitamin B_{12} protection in experimental liver injury to the rat by carbon tetrachloride. *Biochem. J.*, **72**: 384.

Kato, R., Takanaka, A., and Oshima, T. (1969). Effect of vitamin C deficiency on the metabolism of drugs and NADPH-linked electron transport system in liver microsomes. *Jap. J. Pharmacol.*, **19**: 25–33.

Kattamis, C., Zannos-Mariolea, L., Franco, A. P., Liddell, J., Lehmann, H., and Davies, D. (1963). The frequency of atypical pseudocholinesterase in British and Mediterranean populations. *Nature*, **196**: 599.

Keane, W. T., Zavon, M. R., and Witherup, S. H. (1961). Dieldrin poisoning in dogs: Relation to obesity and treatment. *Br. J. Ind. Med.*, **26**: 338.

Keating, J. P., Lell, M. E., Strauss, A. W., Zarkowsky, H., and Smith, G. E. (1973). Infantile methemoglobinemia caused by carrot juice. *New Engl. J. Med.*, **288**: 824.

Keity, A. S. (1972). Hereditary methemoglobinemia with deficiency of NADH-methemoglobin reductase. In *The Metabolic Basis of Inherited Disease*, 3rd ed., J. B. Stanbury, J. B. Wyngaarden, and D. S. Fredrickson (Eds.), McGraw-Hill, New York, pp. 1389–1397.

Kellermann, G., Luyten-Kellermann, M., and Shaw, C. (1973). Genetic variation of aryl hydrocarbon hydroxylase in human lymphocytes. *Am. J. Hum. Genet.*, **25**: 327.

Kellermann, G., Shaw, C. R., and Luyten-Kellermann, M. (1973*a*). Aryl hydrocarbon hydroxylase inducibility and bronchogenic carcinoma. *New Engl. J. Med.*, **289**(18): 934.

Kello, D., and Kostial, I. (1973). The effect of milk diet on lead metabolism. *Environ. Res.*, **6**: 355.

Keys, A. (1975). Coronary heart disease: The global picture. *Atherosclerosis*, **22**: 149–192.

Kilburn, K. H. (1972). Modes of alveolar response to insult. In *Environmental Factors in Respiratory Disease*, D. H. K. Lee (Ed.), Fogarty International Center Proceedings, no. 11. Academic, New York, p. 256.

Kilburn, K. H. (1973). Biological effects of cigarette smoking in the pathogenesis of pulmonary disease. *J. Occup. Med.*, **15**(3): 198–201.

King, R. A. (1949). Vitamin E therapy in Dupuytren's contracture. *J. Bone J. Surg.*, **31B**: 443.

Klein, G. (1976). Immune surveillance and malignant neoplasia. *Third International Symposium on Detection and Prevention of Cancer*, Apr. 26–May 1. New York. Conference 10, no. 48 (abstract).

Kleitman, N., and Jackson, D. P. (1950). Body temperature and performance under different routines. *J. Appl. Physiol.*, **3**: 309.

Klenner, F. R. (Nov., 1955). The role of ascorbic acid in therapeutics. *Tri-State Med. J.*

Klingenberg, M. (1958). Pigment of rat liver microsomes. *Arch. Biochem. Biophys.*, **75**: 376–386.

Knotek, Z., and Schmidt, P. (1964). Pathogenesis, incidence and possibilities of preventing alimentary nitrate methemoglobinemia in infants. *Pediatrics*, **34**: 78.

Knox, W. E. (1972). Phenylketonuria. In *The Metabolic Basis of Inherited Disease*, 3rd ed., J. B. Stanbury, J. B. Wyngaarden, and D. S. Fredrickson (Eds.), McGraw-Hill, New York, pp. 326–337.

Knutsen, C. A., and Brewer, G. J. (1966). The micromethemoglobin reduction test for glucose-6-phosphate dehydrogenase deficiency. *Am. J. Clin. Pathol.*, **45**: 82.

Kobayashi, J. (1957). On geographic relationship between the chemical nature of river water and death rate from apoplexy. *Ber. Ohara Inst. Landwirtsch. Biol.*, **XI**(1): 12.

Kociba, R. J., and Sleight, S. D. (1970). Nitrate toxicosis in ascorbic acid deficient guinea pig. *Toxicol. Appl. Pharmacol.*, **16**: 424.

Koistinen, J. (1975). Studies of selective deficiency of serum IgA and its significance in blood transfusion. Doctoral dissertation, University of Helsinki.

Koistinen, J. (1975a). Selective IgA deficiency in blood donors. *Vox Sang.*, **29**: 1. S. Karger AG, Basel, Switzerland.

Koistinen, J. (Aug., 1975b). Personal communication.

Kolbye, A. C., Mahaffey, K. R., Fiorino, J. A., Corneliussen, P. C., and Jelinek, C. F. (May, 1974). Food exposures to lead. *Environ. Health Perspect.*, p. 65.

Kolesnikov, J. P. (1958). Biological and therapeutic effects of manganese. Kharhov (thesis).

Kornberg, A. (1942). Latent liver disease in persons recovered from catarrhal jaundice and in otherwise normal medical students as revealed by the bilirubin excretion test. *J. Clin. Invest.*, **21**: 299–308.

Kosmider, S., and Misiewicz, A. (1973). Experimental and epidemiological investigations of the effect of nitrogen oxides on lipid metabolism. *Int. Arch. Arbeitsmed.*, **31**: 249–256.

Kouri, R. E., Imblum, R. L., and Prough, R. A. (1976). Measurement of aryl hydrocarbon hydroxylase and NADH-dependent cytochrome C reductase activities in mitogen activated human lymphocytes. *Third International Symposium on Detection and Prevention of Cancer*, Apr. 26–May 1. New York. Symposium 36, no. 510 (abstract).

Kreek, M. J., and Sleisenger, M. H. (1968). Reduction of serum unconjugated bilirubin with phenobarbitone in adult congenital non-haemolytic unconjugated hyper-bilirubinaemia. *Lancet*, 1: 73.

Kretzschmer, C. H., and Ellis, F. (1947). The effect of x-rays on ascorbic acid concentration in plasma and in tissue. *Br. J. Radiol.*, 20: 94–99.

Kubota, J., Lazar, V. A., and Losee, F. (1968). Copper, zinc, cadmium and lead in human blood from 19 locations in the United States. *Arch. Environ. Health*, 16: 788–793.

Kueppers, F., Briscoe, W. A., and Bearn, A. G. (1964). Hereditary deficiency of serum a_1-antitrypsin. *Science*, 146: 1678–1679.

Kueppers, F., Fallat, R., and Larson, R. K. (1969). Obstructive lung disease and alpha$_1$-antitrypsin deficiency gene heterozygosity. *Science*, 165: 899.

Kuratsune, M., Kohchi, S., Horie, A., et al. (1971). Test of alcoholic beverages and ethanol solutions for carcinogenicity and tumor-promoting activity. *Gann*, 62: 395–405.

Laberge, C., and Dallaire, L. (1967). Genetic aspects of tyrosinemia in the Chicoutimi region. *Can. Med. Assoc.*, 97: 1099.

La Du, B. N., and Gjessing, L. R. (1972). Tyrosinosis and tyrosinemia. In *The Metabolic basis of Inherited Disease*, 3rd ed., J. B. Stanbury, J. B. Wyngaarden, and D. S. Fredrickson (Eds.), McGraw-Hill, New York, p. 296.

Lahiri, K. D. (1943). Advancement in the treatment of arsenical intolerance. *Indian J. Vener. Dis. Dermatolog.*, 9: 1–2.

Lambert, P. M. (1970). Smoking, air pollution and bronchitis in Britain. *Lancet*, 1: 853–857.

Lamy, J. N., Namy-Provansal, J., de Russe, J., and Weill, J. D. (1967). Dosage automatique de l'eau oxygéné et application à l'étude cinétique de la catalase. *Bull. Soc. Chim. Biol.*, 49: 1167.

Lane, R. E., and Campbell, A. C. P. (1954). Fatal emphysema in two men making a copper cadmium alloy. *Br. J. Ind. Med.*, 11: 118–122.

Laurell, C. B., and Eriksson, S. (1963). Electrophoretic a_1-globulin pattern of serum in a_1-antitrypsin deficiency. *Scand. J. Clin. Lab. Invest.*, 15: 132.

Lazarow, H. (1954). Relation of glutathione to hormone action and diabetes. In *Glutathione*, S. Colowick (Ed.), Academic, New York.

Leber, T. (1871). Leber hereditare und congenitalangelegte schnerbenleiden. *Albrecht V. Graefe's Arch. Ophthalmol.*, 17, 2 Abt., 249–291 (cited in Wilson, 1965a).

Ledingham, J. M. (1953). Distribution of water, sodium and potassium in heart and skeletal muscle in experimental renal hypertension in rats. *Clin. Sci.*, 12: 337.

Lee, D. H. K. (1970). Nitrates, nitrites and methemoglobinemia. *Environ. Res.*, 3: 484.

Lehmann, H., and Huntsman, R. G. (1972). The hemoglobinopathies. In *The Metabolic Basis of Inherited Disease*, 3rd ed., J. B. Stanbury, J. B. Wyngaarden, and D. S. Fredrickson (Eds.), McGraw-Hill, New York, pp. 1398–1431.

Lehmann, H., and Liddell, J. (1964). Genetical variants of human serum pseudocholines-terase. *Prog. Med. Genet.*, 3: 75.

Lehmann, H., and Liddell, J. (1972). The cholinesterase variants. In *The Metabolic Basis of Inherited Disease*, 3rd ed., J. B. Stanbury, J. B. Wyngaarden, and D. S. Fredrickson (Eds.), McGraw-Hill, New York, p. 1730.

Lehmann, H., and Ryan, E. (1956). The familial incidence of low pseudocholinesterase level. *Lancet*, 2: 124.

Leighton, F., Poole, B., Beaufay, H., Baudhuim, P., Coffey, J. W., Fowler, S., and de Dure, Ch. (1968). The large scale operation of peroxisomes, mitochondria, and lysosomes from the livers of rats injected with trition WR1339. *J. Cell Biol.*, 37: 482.

Lester, R., and Schmid, R. (1964). Bilirubin metabolism. *New Engl. J. Med.*, 270(15): 779.

Lewin, S. (1974). Vitamin C and the common cold. *Chem. Br.*, 10: 25–27.

Lewinsohn, H. C. (1972). The medical surveillance of asbestos workers. *R. Soc. Health J.*, 92: 69.

Lewis, G. P., Jusko, W. J., Coughlin, L. L., and Hartz, S. (1972). Contribution of cigarette smoking to cadmium accumulation in man. *Lancet*, 1: 291–292.

Lewis, R., Gilkeson, M., Jr., and McCaldrin, R. O. (1962). Air pollution and New Orleans asthma. *Pub. Health Rep.*, 77: 947–954.

Lichtman, H. C., and Feldman, F. (1963). In vitro pyrrole and porphyrin synthesis in lead poisoning and iron deficiency. *J. Clin. Invest.*, 42: 830.

Liddell, J., Lehmann, H., and Davies, D. (1963). Harris and Whittaker's pseudocholinesterase variant with increased resistance to fluoride. *Acta Genet.*, 13: 95.

Lieberman, J. (1969). Heterozygous and homozygous $alpha_1$ antitrypsin deficiency in patients with pulmonary emphysema. *N. Engl. J. Med.*, 281: 279–284.

Lieberman, J. (1972). Digestion of antitrypsin deficient lung by leukoproteases. In *Pulmonary Emphysema and Proteolysis*, Charles Mittman (Ed.), Academic, New York, p. 562.

Lieberman, J., and Mittman, C. (May 22, 1972). A new "double-ring" screening test for carriers of a_1-antitrypsin variants. Presented at American Thoracic Society Meeting, Kansas City.

Lieberman, J., Mittman, C., and Schneider, A. S. (1969). Screening for homozygous and heterozygous alpha₁-antitrypsin deficiency. *J. Am. Med. Assoc.*, 210: 2055.

Lieberman, J., and Mohamed, A. G. (1971). Inhibitors and activators of leukocytic proteases in purulent sputum. *J. Lab. Clin. Med.*, 77: 713–727.

Lindemann, R., Gjessing, L. R., Merton, B., Christie Loken, A., and Haluorsen, S. (1970). Amino acid metabolism in fructosemia. *Acta Paediat. Scand.*, 59: 141.

Linderholm, H., and Lundstrom, P. (1969). Endogenous CO production and blood loss at delivery. *Acta Obstet. Gynecol. Scand.*, 48: 362.

Lobeck, C. G. (1972). Cystic fibrosis. In *The Metabolic Basis of Inherited Disease*, 3rd ed., J. B. Stanbury, J. B. Wyngaarden, and D. S. Fredrickson (Eds.), McGraw-Hill, New York, pp. 1605–1626.

Lokietz, H., Dowben, R. M., and Hsia, D. Y. (1963). Studies on the effect of novobiocin and glucuronyl transferase. *Pediatrics*, 32: 47.

Loomis, W. F. (1967). Skin-pigment regulation of vitamin D biosynthesis in man. *Science*, 157: 501–506.

Lougenecker, H. E., Fricke, H. H., and King, C. G. (1940). The effect of organic compounds upon vitamin C synthesis in the rat. *J. Biol. Chem.*, 135: 497–510.

Luce, G. G. (1970). Biological rhythms in psychiatry and medicine. U.S. Dept. HEW, Public Health Service. Pub. no. 2088, pp. 1–15.

Luce, G. G. (1973). *Body Time.* Pantheon, New York.

Ludewig, S., and Chanutin, A. (1950). Distribution of enzymes in the livers of control and x-irradiated rats. *Arch. Biochem.*, **29**: 441.

Lurie, J. B. (1965). Benzene intoxication and vitamin C. *Trans. Assoc. Ind. Med. Off.*, **15**: 78–79.

Lyon, C. J. (1968). Acute porphyria in American Negro. *N.Y. J. Med.*, **68**: 2441.

MacDonald, E. J. (1966). Discussion comment. In *Structure and Control of the Melanocyte*, G. Della Porta and O. Muhlbock (Eds.), Springer-Verlag, New York, p. 317.

MacKenzie, J. B. (1954). Relation between serum tocopherol and hemolysis in hydrogen peroxide of erythrocytes in premature infants. *Pediatrics*, **13**: 346.

MacMahon, B. (1962). Pre-natal x-ray exposure and childhood cancer. *J. Natl. Cancer Inst.*, **28**: 1173–1191.

MacMahon, B., and Hutchison, G. B. (1964). Parental x-ray and childhood cancer: A review. *Rev. Acta Un. Int. Cancer*, **20**: 1172.

Magee, P. N., and Swann, P. F. (1969). Nitroso compounds. *Br. Med. Bull.*, **25**: 240–244.

Maggioni, G., Bottini, E., and Biagi, G. (1960). Contributo alla conoscenza dell'-aminoaciduria qualitativa e quantitativa dell'intossicazione da piomba wel bambino. *Bull. Soc. Ital. Biol. Sper.*, **36**: 193.

Mahaffey, K. R. (1974). Nutritional factors and susceptibility to lead toxicity. *Environ. Health Perspect. Exp.*, no. 7, p. 107.

Mahaffey-Six, K., and Goyer, R. A. (1972). The influence of iron deficiency on tissue content and toxicity of ingested lead in the rat. *J. Lab. Clin. Med.*, **79**: 128.

Marchmont-Robinson, S. W. (1941). Effect of vitamin C on workers exposed to lead dust. *J. Lab. Clin. Med.*, **26**: 1478–1481.

Marier, J. R. (1968). The importance of dietary magnesium with particular reference to humans. National Research Council of Canada, no. 10, 173.

Marier, J. R., and Rose, D. (1966). The fluoride content of some foods and beverages. *J. Food Sci.*, **31**: 941.

Marier, J. R., Rose, D., and Bovlet, M. (1963). Accumulation of skeletal fluoride and its implications. *Arch. Environ. Health*, **6**: 664–671.

Marocco, N., and Rigotti, E. (1962). Kidney protective effect of vitamin C in arsenic poisoning. *Minerva Urol.*, **14**: 207–212.

Marston, R. H. (1954). *Physiology and Biochemistry of the Skin*, S. Rothman (Ed.), University of Chicago Press, Chicago.

Marver, H. S., and Schmid, R. (1968). Biotransformation in the liver: Implications for human disease. *Gastroenterology*, **55**: 282.

Marver, H. S., and Schmid, R. (1972). The porphyrias. In *The Metabolic Basis of Inherited Disease*, 3rd ed., J. B. Stanbury, J. B. Wyngaarden, and D. S. Fredrickson (Eds.), McGraw-Hill, New York, p. 1087.

Marx, J. L. (1975). Thymic hormones: Inducers of T cell maturation. *Science*, **187**: 1183.

Masironi, R. (1969). Trace elements and cardiovascular diseases. *Bull. World Health Organ.*, **40**: 305–312.

Mass. Dept. Pub. Health (1969, 1970). Report of routine chemical and physical analyses of public water supplies.

Mass. Dept. Public Health. (1975). Report of routine chemical and physical analyses of public water supplies.

Masterson, J. W. (Apr. 18, 1975). Chicago Board of Health testimony before the Illinois Pollution Control Board.

Matzen, R. N. (1957). Effects of vitamin C and hydrocortisone on the pulmonary edema produced by ozone in mice. *J. Appl. Physiol.*, **11**: 105.

Maugh, T. H. (1976). The ozone layer: The threat from aerosol cans is real. *Science*, **194**: 170–172.

Mauzerall, D., and Granick, S. (1956). The occurrence and determination of δ-amino levulinic acid and porphobilinogen in urine. *J. Biol. Chem.*, **219**: 435.

Mavin, J. V. (1941). Experimental treatment of acute mercury poisoning of guinea pigs with ascorbic acid. *Rev. Soc. Argent. Biol.*, **17**: 581–586.

Mawhinney, H., and Tomkin, G. H. (1971). Gluten enteropathy associated with selective IgA deficiency. *Lancet*, **2**: 121–124.

McArdle, B. (1940). The serum choline esterase in jaundice and disease of the liver. *Q. J. Med.*, **9**: 875–895.

McChesney, E. W. (1945). Further studies on the detoxication of arphenamines by ascorbic acid. *J. Pharmacol. Exp. Ther.*, **84**: 222–235.

McChesney, E. W., Barlow, O. W., and Kinck, G. H., Jr. (1942). Detoxication of neoarsphenamine by means of various organic acids. *J. Pharmacol. Exp. Ther.*, **80**: 81–92.

McCormick, W. J. (1952). Ascorbic acid as a chemotherapeutic agent. *Arch. Pediatr.*, **69**: 151–159.

McGovern, V. J. (1966). Melanoblastoma in Australia. In *Structure and Control of the Melanocyte*, G. Della Porta and O. Muhlbock (Eds.), Springer-Verlag, New York, pp. 312–315.

McKusick, V. A. (1970). Human genetics. *Ann. Rev. Genet.*, **4**: 1.

McLean, A. E. M. (1970). The effect of protein deficiency and microsomal enzyme induction by DDT and phenobarbitone on the acute toxicity of chloroform and pyrrolizidine alkaloid, retrosine. *Br. J. Exp. Pathol.*, **51**: 317–321.

Mena, I., Horivchi, K., Burke, K., and Cotzias, G. C. (1969). Chronic manganese poisoning: Individual susceptibility and absorption of iron. *Neurology*, **19**: 1000.

Menden, E. E., Elia, V. J., Michael, L. W., and Petering, H. G. (1973). Distribution of cadmium and nickel of tobacco during cigarette smoking. *Environ. Sci. Technol.*, **6**(9): 830–832.

Menzel, C. E. (1968). Rancidity of the red cell: Peroxidation of the red cell lipid. *Am. J. Med. Sci.*, **255**: 341.

Menzel, D. B. (1971). Oxidation of biologically active reducing substances by ozone. *Arch. Environ. Health*, **23**: 149.

Menzel, D. B. (1976). Oxidants and human health. *J. Occup. Med.*, **18**(5): 342–345.

Miller, E. B., Kanabrocki, E. L., Case, L. F., et al. (1967). Non-dialyzable manganese, copper, and sodium in human bile. *J. Nucl. Med.*, **8**: 891–895.

Mills, J. N. (1967). Circadian rhythms and shift workers. *Trans. Soc. Occup. Med.*, **17**: 5.

Ministry of Health. (1954). Mortality and morbidity during the London fog of December, 1952. *Rep. Public Health Relat. Subj.*, no. 95. Her Majesty's Stationery Office, London.

Mitchell, D. G., and Aldous, K. M. (1974). Lead content of food. *Environ. Health Perspect. Exp.*, no. 7, p. 59.

Mitchell, G. E., Jr., Little, C. A., and Skersk, G. (1968). Influence of subacute levels of dietary toxins on carotene disappearance from the rat intestine. *Int. Z. Vitaminforsch.*, **38**: 308 (cited in Shakman, 1974).

Mittler, S. (1958). Protection against death due to ozone poisoning. *Nature*, **181**: 1063.

Mittman, C., and Lieberman, J. (1973). Screening for a_1-antitrypsin deficiency. In *Genetic Polymorphisms and Diseases in Man*. B. Ramot, A. Adam, B. Bonne, R. Goodman, and A. Szeinberg (Eds.), Academic, New York, pp. 185–192.

Mittman, C., Lieberman, J., Marasso, F., and Miranda, A. (1971). Smoking and chronic obstructive lung disease in alpha$_1$-antitrypsin deficiency. *Chest*, **60**: 214.

Mittman, C., Lieberman, J., Miranda, A., and Marasso, F. (1972). Pulmonary disease in intermediate alpha$_1$-antitrypsin deficiency. In *Pulmonary Emphysema and Proteolysis*, C. Mittman (Ed.), Academic, New York, p. 33.

Mokranjac, M., and Petrovic, C. (1964). Vitamin C as an antidote in poisoning by fatal doses of mercury. *C. R. Hebd. Séances Acad. Sci.*, **258**: 1341–1342.

Monier, M. M., and Weiss, R. J. (1952). Increased excretion of dehydroascorbic acid and diketogulonic acids by rats after x-ray irradiation. *Proc. Soc. Exp. Biol. Med.*, **81**: 598–599.

Monsen, E. R., Kuhn, I. N., and Finceh, C. A. (1967). Iron status of menstruating women. *Am. J. Clin. Nutr.*, **20**: 842.

Morgan, A. F. (1959). Nutrition Status U.S.A. Agricultural Experiment Station, Berkeley, Calif., Bulletin 769.

Morris, J. N., Crawford, M. D., and Heady, J. A. (1961). Hardness of local water supplies and mortality from cardiovascular disease. *Lancet*, **1**: 860.

Morrow, A. C., and Motulsky, A. G. (1968). Rapid screening method for the common atypical pseudocholinesterase variant. *J. Lab. Clin. Med.*, **71**: 350.

Morrow, P. E. (1967). Adaptations of the respiratory tract to air pollutants. *Arch. Environ. Health*, **14**: 127–136.

Morrow, P. E. (1975). An evaluation of recent NO_x toxicity data and an attempt to derive an ambient air standard for NO_x by established toxicological procedures. *Environ. Res.*, **10**: 92–112.

Motley, H. L., Smart, R. H., and Leftwich, C. I. (1959). Effect of polluted Los Angeles air (smog) on lung volume measurements. *J. Am. Med. Assoc.*, **71**: 1469–1477.

Mott, P. E., et al. (1965). *Shift work*. The University of Michigan Press, Ann Arbor.

Mottram, J. L. (1945). A diurnal variation in the production of tumours. *J. Pathol. Bacteriol.*, **57**: 265.

Motulsky, A. G. (1973). Screening for sickle cell homoglobinopathy and thalassemia. In *Genetic Polymorphisms and Diseases in Man*, Academic, New York, p. 215.

Motulsky, A. G., and Campbell-Kraut, J. M. (1961). Population genetics of glucose-6-phosphate dehydrogenase deficiency of the red cell. In *Proceedings of Conference on Genetic Polymorphism and Geographic Variations in Disease*, Grune and Stratton, New York, p. 159.

Mountain, J. T. (1963). Detecting hypersusceptibility to toxic substances. *Arch. Environ. Health*, **6**: 357.

Mountain, J. T., Delker, L. L., and Stokinger, H. E. (1953). Studies in vanadium toxicology: I. Reduction in the cystine content of rat hair. *Arch. Ind. Hyg. Occup. Med.*, **8**: 406.

Mountain, J. T., Stockell, F. R., Jr., and Stokinger, H. E. (1955). Studies in vanadium toxicology: III. Fingernail cystine as an early indicator of metabolic changes in vanadium workers. *A.M.A. Arch. Ind. Hyg.*, **12**: 494.

Mountain, J. T., Stockell, F. R., and Stokinger, H. E. (1956). Effects of ingested vanadium on cholesterol and phospholipid metabolism in the rabbit. *Proc. Soc. Exp. Biol.*, **92**: 582–587.

Mountain, J. T., Wagner, W. D., and Stokinger, H. E. (1959). Effects of vanadium on growth, cholesterol metabolism, tissue components in laboratory animals on various diets. *Fed. Proc.*, **18**(pt. 1): 1678.

Mudd, S. H., Irreverre, F., Laster, L. (1967). Sulfite oxidase deficiency in man: Demonstration of the enzyme defect. *Science*, **156**: 1599–1602.

Mueller, G. C., and Miller, J. A. (1948). The metabolism of 4-dimethylaminoazobenzene by rat liver homogenates. *J. Biol. Chem.*, **176**: 535.

Mueller, G. C., and Miller, J. A. (1953). The metabolism of methylated aminoazodyes: II. Oxidative demethylation by rat liver homogenates. *J. Biol. Chem.*, **202**: 579.

Murphy, R. J. F. (1950). The effect of "rice diet" on plasma volume and extracellular fluid space in hypertensive subjects. *J. Clin. Invest.*, **29**: 912.

Muss, D. L. (1962). Relationship between water quality and deaths from cardiovascular disease. *J. Am. Water Works Assoc.*, **54**: 1371.

Naeye, R. L., Mahon, J. K., and Dellinger, W. S. (1971). Effects of smoking on lung structures of Appalachian coal-workers. *Arch. Environ. Health*, **22**: 190–193.

Nalbandian, R. M. (1972). Mass screening programs for sickle cell hemoglobin (editorial). *J. Am. Med. Assoc.*, **221**: 500.

Nandi, M., Gick, H., Slone, D., et al. (1969). Cadmium content of cigarettes. *Lancet*, **2**: 1329–1330.

National Academy of Sciences, National Research Council (1968). Recommended Daily Allowances, 7th ed., publ. 1964. Washington, D.C.

National Academy of Sciences. (1975). Special risk due to inborn error of metabolism. *Principles for Evaluating Chemicals in the Environment.* Washington, D.C., p. 331.

National Academy of Sciences. (1975a). Estimates of increase in skin cancer due to increases in ultraviolet radiation caused by reduced stratospheric ozone, Appendix C. In *Environment Impact of Stratospheric Flight*, pp. 117–221.

Natvig, J. B., Harboe, M., Fausa, O., and Tveit, A. (1971). Family studies in individuals with absence of gamma-A-globulin. *Clin. Exp. Immunol.*, **8**: 229–236.

Neilson, D. R., and Neilson, R. P. (1958). Porphyrin complicated by pregnancy. *West. J. Surg.*, **66**: 133.

Nitrates, nitrites, and nitrosamines. *FDA Fact Sheet*, Jan. 21, 1971.

Nizhegorodov, V. M. (1962). Effects of chronic carbon monoxide poisoning on 24-hour vitamin C requirements in animals. *Zdravo-Okhr.* **8**: 50–53.

Nutrition Canada. (1973). National Survey (Ed.). Information Canada. Ottawa, Ontario, Canada.

Nyhan, W. L. (1961). Toxicity of drugs in the neonatal period. *J. Pediatr.*, **59**(1): 1.

O'Beirn, S. F., Judge, P., Urbach, F., MacCon, C. F., and Martin, F. (1970). The prevalence of skin cancer in County Galway, Ireland. In *Proceedings of the 6th National Cancer Conference*, Lippincott, Philadelphia, pp. 489–500.

O'Brien, R. D. (1967). *Insecticides: Action and Metabolism*, Academic, New York, pp. 15–21.

Ochs, S. (1971). General properties of nerve. In *Physiology*, E. E. Selkurt (Ed.), Little, Brown, Boston, p. 39.

Ochs, S. (1971*a*). Receptors and effectors: A. Sensation and neuromuscular transmission. In *Physiology*, E. E. Selkurt (Ed.), Little, Brown, Boston.

Oehme, F. M. (1969). A comparative study of the bio-transformation and excretion of phenol. Ph.D. dissertation, University of Missouri.

Oettle, A. G. (1966). Epidemiology of melanomas in South Africa. In *Structure and Control of the Melanocyte*, G. Della Porta and O. Muhlbock (Eds.), Springer-Verlag, New York, pp. 292–307.

Oliver, T. K., Jr. (1958). Chronic vitamin A intoxication: Report of a case in an older child and review of the literature. *Am. J. Dis. Child.*, **95**: 57.

Olsen, K. (1976). Review of federal toxic substances legislation. New England Air Pollution Control Association Meeting, Nov. 7–9, 1976, Cambridge, Mass.

Omura, T., and Sato, R. (1964). The carbon monoxide-binding pigment of liver microsomes: I. Evidence for its hemeprotein nature. *J. Biol. Chem.*, **239**: 2370–2378.

Omura, T., and Sato, R. (1964*a*). The carbon monoxide-binding pigment of liver microsomes: II. Solubilization, purification and properties. *J. Biol. Chem.*, **239**: 2379–2385.

Orzales, M. M., Kohner, D., Cook, C. D., and Shwachman, H. (1963). Anamnesis, sweat electrolyte and pulmonary function studies in parents of patients with cystic fibrosis of the pancreas. *Acta Paediatr. Scand.*, **52**: 267.

OSHAct of 1970 (Public Law 91-596), Section 6(6) (5).

Oski, F. A., and Growney, P. M. (1965). A simple micromethod for the detection of erythrocyte glucose-6-phosphate dehydrogenase deficiency (G-6-PD spot test). *Blood*, **20**: 591.

Oster, H. L., et al. (1953). X-Irradiation on ascorbic acid of rat tissues. *Proc. Soc. Exp. Biol. Med.*, **84**: 470–473.

Ostwald, R., and Briggs, G. M. (1966). Toxicity of the vitamins. In *Toxicants Occurring Nationally in Foods*. National Research Council, National Academy of Sciences, Washington, D. C.

Ott, M. G., Holder, B. B., and Gordon, H. L. (1974). Respiratory cancer and occupational exposure to arsenicals. *Arch. Environ. Health*, **29**: 250.

Otten, J., and Vis, H. L. (1968). Acute reversible renal tubulus dysfunction following intoxication with methyl-3-chromone. *J. Pediatr.*, **73**: 422.

Owens, D., and Evans, J. (1975). Population studies on Gilbert's syndrome. *J. Med. Genet.*, **12**: 152–156.

Pagnotto, L. D., and Epstein, S. S. (1969). Protection by anti-oxidants against ozone toxicity in mice. *Experientia*, **25**: 703.

Pai, M. K. R., Davidson, M., Dedritis, I., and Zipursky, A. (1974). Selective IgA deficiency in Rh-negative women. *Vox Sang.*, **27**: 87–91.

P'an, A. Y. S., Beland, J., and Zygmont, J. (1972). Ozone-induced arterial lesions. *Arch. Environ. Health*, **24**: 229–232.

Paniker, N. V. (1975). Red cell enzymes. *CRC Crit. Rev. Clin. Lab. Sci.*, April, p. 469.

Parizek, J. (1957). The destructive effect of cadmium ion upon the testicular tissue and its prevention by zinc. *J. Endocr.*, **15**: 56.

Parizek, J. (1965). The peculiar toxicity of cadmium during pregnancy: An experimental "toxemia of pregnancy" induced by cadmium salts. *J. Reprod. Fertil.*, **9**: 11.

Parizek, J., Ostadalova, I., Benes, I., and Pitha, J. (1968). The effects of subcutaneous injection of cadmium salts on the ovaries of adult rats in persistent oestrus. *J. Reprod. Fertil.*, **17**: 559.

Parizek, J., Ostadalova, I., Benes, I., and Babicky, A. (1968a). Pregnancy and trace elements: The protective effect of compounds of an essential trace element—selenium—against the peculiar toxic effects of cadmium during pregnancy. *J. Reprod. Fertil.*, **16**: 507.

Parizek, J., Ostadalova, I., Kalouskova, J., Babicky, A., and Benes, I. (1971). The detoxifying effects of selenium: Interrelations between compounds of selenium and certain metals. In *Newer Trace Elements in Nutrition*, W. Mertz and W. E. Cornatzer (Eds.), Marcel Dekker, New York, p. 85.

Patty, R. A. (1962). *Industrial Hygiene and Toxicology*, Vol. II, Interscience, New York, p. 932.

Peacock, P. R., and Spence, J. B. (1967). Incidence of lung tumours in LX mice exposed to (1) free radicals; (2) SO_2. *Br. J. Cancer*, **21**: 606–618.

Pearlman, M. E., Finklea, J. F., Creason, J. P., Shy, C. M., Young, M. M., and Horton, R. J. M. (1971). Nitrogen dioxide and lower respiratory tract illness. *Pediatrics*, **47**(2): 391–398.

Pearson, H. A., O'Brien, R. T., and McIntosh, S. (1973). Screening for thalassemia trait by electronic measurement of M.C.V. *New Engl. J. Med.*, **288**: 1129.

Pelkonen, O., Kaltiala, E. H., Larmi, T. K. and Karki, N. T. (1973). Comparison of activities of drug-metabolizing enzymes in human fetal and adult livers. *Clin. Pharm. Therap.*, **14**: 840–846.

Pelkonen, O., Vorne, M., Arvela, P., Jouppila, P., and Karki, N. T. (1971). Drug metabolizing enzymes in human fetal liver and placenta in early pregnancy. *Scand. J. Clin. Lab. Invest.*, Suppl. 116. **27**: 7.

Pelletier, O. (1968). Smoking and vitamin C levels in humans. *Am. J. Chem. Nutr.*, **21**: 1259–1267.

Pelletier, O. (1970). Vitamin C status of cigarette smokers and non-smokers. *Am. J. Clin. Nutr.*, **23**: 520–524.

Pelletier, O. (1975). Vitamin C and cigarette smokers. In *Ann. N.Y. Acad. Sci.*, Second Conference on Vitamin C, C. G. King and J. J. Burns (Eds.). **258**: 156–167.

Penny, R., Thompson, R. G., Polmar, S. H., and Schultz, R. B. (1971). Pancreatitis, malabsorption and IgA deficiency in a child with diabetes. *J. Pediatr.*, **78**: 512–516.

Perlroth, M. G., Marver, H. S., and Tschudy, D. P. (1965). Oral contraceptive agents and the management of acute intermittent porphyria. *J. Am. Med. Assoc.*, **194**: 1037.

Perutz, M. F., and Mitchison, J. M. (1950). State of haemoglobin in sickle-cell anemia. *Nature*, **166**: 677.

Peters, J. M., and Murphy, R. L. H. (1970). Pulmonary toxicity of isocyanates. *Ann. Intern. Med.*, **73**(4): 654–655.

Peterson, N. J., Samuels, L. D., Lucas, H. F., and Abraham, S. P. (1966). An epidemiologic approach to low-level radium-226 exposure. *Public Health Rep.*, **81**(9): 805..

Petrie, S. J., and Mooney, J. B. (1962). Porphyria with the complication of pregnancy. *Am. J. Obstet. Gynecol.*, **83**: 264.

Petuknov, N. I., and Ivanov, A. C. (1970). Investigation of certain psychophysiological reactions in children suffering from methemoglobinemia due to nitrates in water. *Hyg. Sanit.*, **35**: 29.

Philpott, M. G. (1975). Infant feeds and softened water. *Lancet*, **1**(7921): 1378.

Piedrabuena, L. (1970). Experiments on the production of hypervitaminosis E in birds by giving DL-a-tocopherol acetate. *Nutr. Abstr. Rev.*, **40**: 48.

Pillemer, L., Seifter, J., Kuehn, A. O., and Ecker, E. E. (1940). Vitamin C in chronic lead poisoning. *Am. J. Med. Sci.*, **200**(3): 322–327.

Pirozzi, D. J., Gross, P. R., and Samitz, M. H. (1968). The effect of ascorbic acid on chrome ulcers in guinea pigs. *Arch. Environ. Health*, **17**: 178–180.

Pollack, S., James, N., George, R. C., Kaufman, R. M., and Crisby, W. H. (1965). The absorption of non-ferrous metal in iron deficiency. *J. Clin. Invest.*, **44**: 1470

Polmar, S. H., Waldmann, T. A., Balestra, S. T., Jost, M. C., and Terry, W. D. (1972). Immunoglobulin E in immunologic deficiency disease: I. Relation of IgE and IgA to respiratory tract disease in isolated IgE deficiency, IgA deficiency and ataxia telangiectasis. *J. Clin. Invest.*, **51**: 326–330.

Potter, S. D., and Matrone, G. (1973). Effect of selenite on the toxicity and retention of dietary methyl mercury and mercuric chloride. *Fed. Proc.*, **32**: 929.

Powell, L. W., Hemingway, E., Billing, B. H., and Sherlock, S. (1967). Idiopathic unconjugated hyperbilirubinemia (Gilbert's syndrome): A study of 42 families. *New Engl. J. Med.*, **277**: 1108.

Powers, E. L., Jr. (Chairman). (1963). Peroxides in radiobiology: A synthesis. *Radiat. Res. Suppl.*, **3**: 270.

Public Health Service. (1962). Drinking water standards. PHS pub. 956. Government Printing Office.

Public Health Service. (1968). The health consequences of smoking: A Public Health Service review, 1967. Washington, D.C.

Radford, E. P. (1976). Carbon monoxide and human health. *Jour. Occup. Med.*, **18**(5): 310–315.

Rakitzis, E. T. (1964). Test for glucose-6-phosphate dehydrogenase deficiency. *Lancet*, **ii**: 1182.

Ramos, A., Silverberg, M., and Stern, L. (1966). Pregnanediols and neonatal hyperbilirubinemia. *Am. J. Dis. Child.*, **111**: 353.

Rane, A., and Ackermann, E. (1971). Evidence for drug metabolism in the human fetal liver: Studies in different cell fractions. *Acta Pharmacol. Toxicol.*, Suppl. 4. **29**: 84.

Rane, A., and Ackermann, E. (1972). Metabolism of ethylmorphine and aniline in human fetal liver. *Clin. Pharmacol. Ther.*, **13**: 663–670.

Rane, A., Von Bahr, C., Orrenius, S., and Sjogvist, F. (1973). Drug metabolism in the human fetus. In *Fetal Pharmacology*, Raven, New York, pp. 287–303.

Rasmussen, R. A. (1976). Surface ozone observation in rural and remote areas. *Jour. Occup. Med.*, **18**(5): 346.

Recknagel, R. O. (1967). Carbon tetrachloride hepatotoxicity. *Pharmacol. Rev.*, **19**: 145–208.

Remmer, H. (1965). The fate of drugs in the organism. *Rev. Pharmacol.*, **5**: 405.

Resnick, H., Lapp, N. L., Keith, W., and Morgan, C. (1971). Serum trypsin inhibitor concentrations in coal miners with respiratory symptoms. *J. Am. Med. Assoc.*, **215**: 1101.

Rimington, C., Magnus, I. A., Ryan, E. A., and Cripps, D. J. (1967). Porphyria and photosensitivity. *Q. J. Med.*, **36**: 29.

Robbins, K. C. (1961). Enzymatic omega oxidation of fatty acids. *Fed. Proc.*, **20**: 272.

Robertson, J. S. (1968). Mortality and hardness of water. *Lancet*, **11**: 348.

Robertson, J. S. (1975). Water sodium: The problem of the bottle fed neonate. Paper presented at the Water Research Centre, Nov. 4–6.

Roehm, J. N., Hadley, J. A., and Menzel, D. B. (1971). Antioxidants vs. lung disease. *Arch. Int. Med.*, **128**: 88.

Roehm, J. N., Hadley, J. A., and Menzel, D. B. (1971a). Oxidation of unsaturated fatty acids by ozone and nitrogen dioxide. *Arch. Environ. Health.*, **23**: 142.

Roehm, J. N., Hadley, J. A., and Menzel, D. B. (1972). The influence of vitamin E on the lung fatty acids of rats exposed to ozone. *Arch. Environ. Health*, **24**: 237.

Roels, O. A. (1966). Present knowledge of vitamin A. *Nutr. Rev.*, **25**(5): 129.

Rose, C. S., and Gyorgy, P. (1950). Hemolysis with alloxan and alloxan-like compounds, and the protective action of tocopherol. *Blood*, **5**: 1062.

Rosenlund, M. L., Kim, H. K., and Kritchevsky, D. (1974). Essential fatty acids in cystic fibrosis. *Nature*, **251**: 719.

Rothman, K. J. (1975). Alcohol. In *Persons at High Risk of Cancer: An Approach to Cancer Etiology and Control*, J. F. Fraumeni, Jr. (Ed.), Academic, New York, pp. 139–150.

Rothman, K. J., and Keller, A. Z. (1972). The effect of exposure pressure to alcohol and tobacco on the risk of cancer of the mouth and pharynx. *J. Chron. Dis.*, **25**: 711–716.

Rotruck, J. T., Hoekstra, W. G., and Pope, A. L. (1971). Glucose-dependent protection by dietary selenium against hemolysis of rat erythrocytes in vitro. *Nature New Biol.*, **231**: 223.

Rotruck, J. T., Pope, A. L., Ganther, H. E., and Hoekstra, W. G. (1972). Prevention of oxidative damage to rat erythrocytes by dietary selenium. *J. Nutr.*, **102**: 689.

Rouiller, C. (1964). Experimental toxic injury of the liver. In *The Liver*, Vol. 2, C. Rouiller (Ed.), Academic, New York, pp. 335–476.

Rowe, N. H., and Gorlin, R. J. (1959). The effect of vitamin A deficiency upon experimental oral carcinogenesis. *J. Dent. Res.*, **38**: 72.

Rupniewska, Z. N. (1965). Duration of smoking and content of ascorbic acid in the body. *Pol. Tyg. Lek.*, **20**: 1069–1071.

Rusev, G., Radev, T., Belikonski, I., and Petkov, B. (1960). Radiochuvstvitelnost na khipokata laznite morski svincheta s ogled na znachenieto na vodorodniya prekis za patogenezata na ostrata lycheva bolest. *Izv. Inst. Stravitclna Patol. Domashnite Zhivotn.*, **8**: 119–129.

Russell, E. L. (1969). Sodium imbalance in drinking water. *J.A.W.W.A.*, **62**(2): 102–105.

Saffiotti, U., Montesano, R., Sellakumar, A. R., and Borg, S. A. (1967). Experimental cancer of the lung: Inhibition by vitamin A of the induction of tracheobronchial squamous metaphasia and squamous cell tumors. *Cancer*, **20**(5): 857.

Saint, E. G., and Curnow, D. H. (1962). Porphyria in western Australia. *Lancet*, **1**: 133.

Saita, G., and Moreo, L. (1959). Talassemia eo emopatie professional: Nota 1-thalassemia e benzolismo cronico. (Thalassemia and occupational blood disease: I. Thalassemia and chronic benzol poisoning). *Med. Lav.*, **50**: 25.

Samitz, M. H., Shrager, J. D., and Katz, S. A. (1962). Studies on the prevention of injurious effects of chromates in industry. *Ind. Med. Surg.*, **31**: 427–432.

Samitz, M. H. (1970). Ascorbic acid in the prevention and treatment of toxic effects of chromates. *Acta Derm. Vener.*, **50**: 59.

Samitz, M. H., Scheiner, D. M., and Katz, S. A. (1968). Ascorbic acid in the prevention of chrome dermatitis. *Arch. Environ. Health*, **17**: 44–45.

Sass, M. D., Caruso, C. J., and Axelrod, D. R. (1966). Rapid screening for D-glucose-6-phosphate: NADP oxidoreductase deficiency with methylene blue. *J. Lab. Clin. Med.*, **68**: 156.

Savilahti, E. (1973). IgA deficiency in children. Immunoglobulin containing cells in the intestinal mucosa, immunoglobulins in secretions and serum IgA levels. *Clin. Exp. Immunol.*, **13**: 395–406.

Scheuch, D., and Kutscher, H. (1963). Eine einfache Bestimmung der Glucose-6-Phosphat dehydrogenase als tupfelprobe zur Verwendung als Siebtest. *Z. Med. Lab. Tech.*, **3**: 22.

Schimmel, H., and Murawski, T. J. (1976). The relationship of air pollution to mortality. *J. Occup. Med.*, **18**(5): 316–333.

Schlegel, J. U., Pipkin, G. E., Nishimura, R., and Duke, G. A. (1969). Studies on the etiology and prevention of bladder carcinoma. *J. Urol.*, **101**: 317–324.

Schmid, R. (1960). Cutaneous porphyria in Turkey. *New Engl. J. Med.*, **263**: 397.

Schneider, J. A., Rosenbloom, F. M., Bradley, K. H., and Seegmiller, J. E. (1967). Increased free-cystine contents of fibroblasts cultured from patients with cystinosis. *Biochem. Biophys. Res. Commun.*, **29**: 527.

Schneider, J. A., and Seegmiller, J. E. (1972). Cystinosis and the Fanconi syndrome. In *The Metabolic Basis of Inherited Disease*, 3rd ed., J. B. Stanbury, J. B. Wyngaarden, and D. S. Fredrickson (Eds.), McGraw-Hill, New York, p. 1581.

Schneider, J. A., Wong, V., Bradley, K. H., and Seegmiller, J. E. (1968). Biochemical comparisons of the adult and childhood forms of cystinosis. *New Engl. J. Med.*, **279**: 1253.

Schrenk, H. H., Heimann, H., Clayton, G. D., Gefafer, W. M., and Wexler, H. (1949). Air pollution in Denora, Pa. *Public Health Bull.*, no. 306. Federal Security Agency, PHS, Division of Industrial Hygiene, Washington.

Schroeder, H. A. (1960). Relation between mortality from cardiovascular disease and tread water supplies. *J. Am. Med. Assoc.*, **172**: 1902.

Schroeder, H. A. (1965). Cadmium as a factor in hypertension. *J. Chron. Dis.*, **18**: 647.

Schroeder, H. A. (1966). Municipal drinking water and cardiovascular death rates. *J. Am. Med. Assoc.*, **195**: 81–85.

Schroeder, H. A. (1967). Essential trace metals in man: Zinc, relation to environmental cadmium. *J. Chron. Dis.*, **20**: 179.

Schroeder, H. A. (1974). The role of trace elements in cardiovascular diseases. *Med. Clin. North America*, **58**(2): 381–396.

Schroeder, H. A., and Balassa, J. J. (1965). Influence of chromium, cadmium and lead on rat aortic lipids and circulating cholesterol. *Am. J. Physiol.*, **209**: 433.

Schroeder, H. A., Balassa, J. J., Gibson, F. S., et al. (1961). Abnormal trace metals in man: Lead. *J. Chron. Dis.*, **14**: 408–425.

Schroeder, H. A., and Buckman, J. (1967). Cadmium hypertension: Its reversals in rats by a zinc chelate. *Arch. Environ. Health*, **14**: 693.

Schroeder, H. A., Frost, D. V., and Balassa, J. J. (1970). Essential trace metals in man: Selenium. *J. Chron. Dis.*, **23**: 227.

Schultze, R. (1974). Increase of carcinogenic ultraviolet radiation due to reduction in ozone concentration in the atmosphere. *Proc. Int. Conf. Structure, Composition and General Circulation of the Upper and Lower Atmospheres and Possible Anthropogenic Perturbations*. Atmospheric Environment Service, Ontario, Canada. **1**: 479–493.

Schulz, J., and Smith, N. J. (1958). A quantitative study of the absorption of food iron in infants and children. *Am. J. Dis. Child.*, **95**: 109.

Schwartz, S., Keprios, M., and Schmid, R. (1952). Experimental porphyria: II. Type produced by lead, phenylhydrazine and light. *Proc. Soc. Exp. Biol. Med.*, **79**: 463.

Schwetz, B. A., Leong, B. K. J., and Gehring, P. J. (1974). Embryo and fetotoxicity of inhaled chloroform in rats. *Toxicol. Appl. Pharmacol.*, **28**: 442–451.

Science News. (Apr. 26, 1976). Chemicals found in cities' water, p. 269.

Science News. (May 13, 1976). Vitamin A and cancer prevention, p. 165.

Scotto, J., Kopf, A. W., and Urbach, F. (1974). Non-melanoma skin cancer among caucasians in four areas of the United States. *Cancer*, **34**: 1333–1338.

Seelig, M. S. (1964). The requirement of magnesium by the normal adult. *Am. J. Clin. Nutr.*, **14**: 342.

Selikoff, I. J. (Ed.). (1972). Polychlorinated biphenyl-environmental impact: A review of the panel on hazardous trace substances. *Environ. Res.*, **5**: 249.

Selikoff, I. J., Churg, J., and Hammond, E. C. (1964). Asbestos exposure and neoplasia. *J. Am. Med. Assoc.*, **188**: 22–26.

Selikoff, I. J., and Hammond, E. C. (1975). Multiple risk factors in environmental cancer. In *Persons At High Risk of Cancer: An Approach to Cancer Etiology and Control*, J. F. Fraumeni, Jr. (Ed.), Academic, New York, p. 467.

Selikoff, I. J., Hammond, E. C., and Churg, J. (1968). Asbestos exposure, smoking and neoplasia. *J. Am. Med. Assoc.*, **204**: 106–112.

Selikoff, I. J., Hammond, E. C., and Seidman, H. (1973). Cancer risk of insulation workers in the United States. In *Biological Effects of Asbestos*, International Agency for Research in Cancer, Lyon, pp. 209–216.

Selkurt, E. E. (1971). Renal function. In *Physiology*, E. E. Selkurt (Ed.), Little, Brown, Boston, p. 487.

Shakman, R. A. (1974). Nutritional influences on the toxicity of environmental pollutants. *Arch. Environ. Health*, **28**: 105.

Shamberger, R. J. (1970). Relationship of selenium to cancer: Inhibitory effect of selenium on carcinogenesis. *J. Natl. Cancer Inst.*, **44**: 931–936.

Shamberger, R. J., and Willis, C. E. (1971). Selenium distribution and human cancer mortality. *CRC Crit. Rev. Clin. Lab. Sci.*, **2**: 211–221.

Shapiro, B., et al. (1968). Ascorbic acid protection against inactivation of lysozyme and aldolase by ionizing radiation. *U.S. Air Force School of Aerospace Med.* SAM-TR-65-71: 1–3.

Shapiro, B., and Kollman, G. (1967). Protection by ascorbic acid against radiation damage in vitro. *J. Albert Einstein Med. Cent.*, **15**: 63–70.

Shapiro, M. P., Keen, P., Cohen, L., and Murray, J. F. (1953). Skin cancer in the South African Bantu. *Br. J. Cancer*, **7**: 45.

Shapiro, R., DiFate, V., and Welcher, M. (1974). Deamination of cytosine derivative by bisulfite: Mechanism of the reaction. *J. Am. Chem. Soc.*, **96**: 906–912.

Shearer, L. A., Goldsmith, J. R., Young, C., Kearns, O. A., and Tamplin, B. (1971). Methemoglobin levels in infants in an area with high nitrate water supply. Presented at the 99th Annual Meeting of the American Public Health Association, Minneapolis, Minn., October, 1971.

Shield, A. M. (1899). A remarkable case of multiple growths of the skin caused by exposure to the sun. *Lancet*, **1**: 22–23.

Short, C. R., and Davies, L. E. (1970). Perinatal development of drug-metabolizing activity in swine. *J. Pharmacol. Exp. Ther.*, **174**: 185–196.

Shy, C. M., Creason, J. P., Pearlman, M. E., McClain, K. E., Benson, F. B., and Young, M. M. (1970). The Chattanooga school children study: Effects of community exposure to nitrogen dioxide: I. Methods, description of pollutant exposure and results of ventilatory function testing. *J. Air Pollut. Control Assoc.*, **20**(8): 539–545.

Shy, C. M., Creason, J. P., Pearlman, M. E., McClain, K. E., Benson, F. B., and Young, M. M. (1970a). The Chattanooga school children study: Effects of community exposure to nitrogen dioxide: II. Incidence of acute respiratory illness. *J. Air Pollut. Control Assoc.*, **20**(9): 582–588.

Silverstone, H., and Gordon, D. (1966). Regional studies in skin cancer, 2nd report: Wet tropical and sub-tropical coast of Queensland. *Med. J. Aust.*, **2**: 733–740.

Silvestroni, E., and Bianco, I. (1959). The distribution of the microcythaemias or thalassaemias in Italy: Some aspects of the haematological and haemoglobinic picture in these haemopathies. In *Abnormal Haemoglobins—A Symposium*, J. N. P. Jonxis and J. F. Dolafresnaye (Eds.), Blackwell, Oxford, p. 242.

Simon, C. (1966). Nitrite poisoning from spinach. *Lancet*, **1**: 872.

Simpson, N. E., and Kalow, W. (1965). Comparisons of two methods for typing of serum cholinesterase and prevalence of its variants in Brazilian population. *Am. J. Hum. Genet.*, **17**: 156.

Six, K. M., and Goyer, R. A. (1970). Experimental enhancement of lead toxicity by low dietary calcium. *J. Lab. Clin. Med.*, **76**: 933.

Sjöstrand, T. (1949). Endogenous formation of carbon monoxide in man under normal and pathological conditions. *Scand. J. Clin. Lab. Invest.*, **1**: 201–214.

Smidt, V., and von Nieding, G. (1974). Airway resistance after inhalation of 1–10 ppm NO_2 and 20–30 ppm with and without previous treatment with broncholytic drugs. *Bull. Physiol. Pathol. Resp.*, **10**: 594.

Smit, J. W., and Hofstede, D. (1966). Vitamin A poisoning in adults. *Ned. Tijdschr. Geneeskd.*, **110**: 10.

Smith, B. A. (1974). Feeding overstrength cows' milk to babies. *Br. Med. J.*, **4**(5947): 741–742.

Smith, E. W., and Conley, C. L. (1955). Sicklemia and infarction of the spleen during aerial flight. *Bull. Johns Hopkins Hosp.*, **96**: 35.

Smith, J. M. (1972). A simple screening test for antitrypsin deficiency. *Ann. Rev. Respir. Dis.*, **105**: 851.

Smith, P. M., Middleton, J. E., and Williams, R. (1967). Studies on the familial incidence and clinical history of patients with chronic unconjugated hyperbilirubinaemia. *Gut*, **8**: 449.

Smith, R. L., and Williams, R. T. (1966). Implication of the conjugation of drugs and other exogenous compounds. In *Glucuronic Acid*, G. J. Dutton (Ed.), Academic, New York, p. 457.

Sobel, A. E., and Burger, M. (1955). Calcification: XIII. The influence of calcium, phosphorus, and vitamin D on the removal of lead from bone. *J. Biol. Chem.*, **212**: 105.

Sobel, A. E., Gawron, O., and Kramer, B. (1938). Influence of vitamin D in experimental lead poisoning. *Proc. Soc. Exp. Biol. Med.*, **38**: 433.

Sorensen, A. J. (1976). Univ. Illinois, School of Public Health (personal communication).

Sparrow, A. H., and Schairer, L. A. (1974). Mutagenic response of *Tradescantia* to treatment with X-rays, EMS, DBE, ozone, SO_2, N_2O and several insecticides, abstracted. *Mutat. Res.*, **21**: 445.

Spock, A., Heick, H. M. C., Cress, H., and Logan, W. S. (1967). Abnormal serum factor in patients with cystic fibrosis of the pancreas. *Pediatr. Res.*, 1: 173.

Stanbury, J. B., Wyngaarden, J. B., and Fredrickson, D. S. (Eds.). (1972). *The Metabolic Basis of Inherited Disease*, McGraw-Hill Book Co., New York, p. 1778.

Stecher, P. G. (Ed.). (1968). *The Merck Index*, 8th ed., Merck and Co., Rahway, N.J., p. 810.

Stein, J. A., Tschudy, D. P., Marver, H. S., Berard, C. W., Ziegel, R. F., Recheigl, M., Jr., and Collens, A. (1966). Acute intermittent porphyria: New morphologic and biochemical findings. *Am. J. Med.*, 41: 149.

Stein, W. H., Bearn, A. G., and Moore, S. (1954). The amino acid content of the blood and urine in Wilson's disease. *J. Clin. Invest.*, 33: 410.

Stenback, F. (1969). The tumorigenic effect of ethanol. *Acta Pathol. Microbiol. Scand.*, 77: 325–326.

Stephens, R. J., Freedman, G., Crane, S. C., and Furiosi, N. J. (1971). Ultrastructural changes in the terminal bronchiole of the rat during continuous low level exposure to nitrogen dioxide. *Exp. Mol. Pathol.*, 14: 1–19.

Sternlieb, I., Morell, A. G., Baver, C. D., Combes, B., Sternberg, S., and Scheinberg, I. H. (1961). Detection of the heterozygous carrier of the Wilson's disease gene. *J. Clin. Invest.*, 40: 707.

Stewart, A., and Kneale, G. W. (1968). Changes in the cancer risk associated with obstetric radiography. *Lancet*, 1: 104–107.

Stewart, A., and Kneale, G. W. (1970). Radiation dose effects in relation to obstetric x-rays and childhood cancer. *Lancet*, 1(7658): 1185–1188.

Stewart, A., Webb, J., and Hewitt, D. (1958). A survey of childhood malignancies. *Br. Med. J.*, 1: 1495–1508.

Stewart, R. D. (1975). The effects of carbon monoxide on humans. *Ann. Rev. Pharm.*, 15: 409–423.

Stewart, R. D. (1976). The effect of carbon monoxide on humans. *J. Occup. Med.*, 18(5): 304–309.

Stewart, R. D., Baretta, E. D., Platte, L. R., Stewart, E. R., Kalbfeisch, J. H., van Yserloo, B., and Rimm, A. A. (1974). Carboxyhemoglobin levels in American blood donors. *J. Am. Med. Assoc.*, 229: 1187–1195.

Stjernswärd, J. (1966). Effect of noncarcinogenic and carcinogenic hydrocarbons on antibody-forming cells measured at the cellular level *in vitro. J. Natl. Cancer Inst.*, 36: 1189–1195.

Stjernswärd, J. (1969). Immunosuppression by carcinogens. *Antibiot. Chemother.* (Basel), 15: 213–233.

Stokinger, H. E. (1962). Effects of air pollution on animals. In: Stern, A. C. (Ed.). *Air Pollution* vol. 1. Academic Press, N.Y., p. 303.

Stokinger, H. E. (1972). Concepts of thresholds in standard setting. *Arch. Environ. Health* 25: 155.

Stokinger, H. E., and Mountain, J. T. (1963). Test for hyper-susceptibility to hemolytic chemicals. *Arch. Environ. Health*, 6: 57.

Stokinger, H. E., and Scheel, L. D. (1973). Hypersusceptibility and genetic problems in occupational medicine: A consensus report. *J. Occup. Med.*, 15: 564–573.

Stone, I. (1972). *The Healing Factor*. Grosset and Dunlap, N.Y., p. 152.

Strauss, L. H., and Scheer, P. (1939). Effect of nicotine on vitamin C metabolism. *Int. Z. Vitaminforsch.*, 9: 39–48.

Street, J. C., and Chadwick, R. W. (1975). Ascorbic acid requirements and metabolism in relationship to organochlorine pesticides. *Ann. N.Y. Acad. Sci.*, **258**: 132–143.

Strittmatter, P. (1965). Protein and coenzyme interactions in the NADH–cytochrome b_5 reductase system. *Fed. Proc.*, **24**: 1156–1163.

Strughold, H. (1952). Physiological day-night cycle in global flights. *J. Aviat. Med.*, **23**: 464.

Subramanian, W., Nandi, B. K., Majumder, A. K., and Chatterjee, I. B. (1973). Role of L-ascorbic acid on detoxification of histamine. *Biochem. Pharmac.*, **22**: 671.

Summers, G. A., and Drake, J. W. (1971). Bisulfite mutagenesis in bacteriophage T4. *Genetics*, **68**: 603–607.

Sutherland, J. M., and Keller, W. H. (1961). Novobiocin and neonatal hyperbilirubinemia. *Am. J. Dis. Child.*, **101**: 447.

Swarm, R. (1971). Congenital hyperbilirubinemia in the rat: An animal model for the study of hyperbilirubinemia. In *Animal Models for Biomedical Research*, IV, National Academy of Sciences, Washington, D.C., pp. 149–160.

Swarup, S., Ghosh, S. K., and Chatterjea, J. B. (1960). Glutathione stability test in hemoglobin E-thalassemia disease. *Nature*, **188**(4745): 153.

Sweeney, G. D. (1963). Patterns of porphyrin excretion in South African patients. *S. Afr. J. Lab. Clin. Med.*, **9**: 182.

Szakal, A. K., and Hanna, M. G. (1972). Immune suppression and carcinogenesis in hamsters during topical application of 7,12-dimethylbenz(a)anthracene. Presented at conference on Immunology of Carcinogenesis at Oak Ridge National Laboratory at Gatlinburg, Tenn., May 8–11.

Szeinberg, A. (1973). Screening for susceptibility to drug reactions: Cholinesterase mutants. In *Genetic Polymorphisms and Diseases in Man*, Academic, New York, p. 229.

Szeinberg, A., deVries, A., Pinkhas, J., Djaldetti, M., and Ezra, R. (1963). A dual hereditary red blood cell defect in one family: Hypocatalasemia and glucose-6-phosphate dehydrogenase deficiency. *Acta Genet. Med. Gemellol.*, **10**: 247.

Szeinberg, A., Pipano, S., Assa, M., Medale, J. H., and Neufeld, H. N. (1972). High frequency of typical pseudocholinesterase gene among Iraqi and Iranian Jews. *Clin. Genet.*, **3**: 123.

Tabershaw, I. R. (1976). Oxides of sulfur. *J. Occup. Med.*, **18**(5): 360–361.

Tabershaw, I. R., and Cooper, W. C. (1966). Sequelae of acute organic phosphate poisoning. *Jour. Occup. Med.*, **8**: 5–20.

Takahara, S. (1967). Acatalasemia. *Asian Med. J.*, **10**: 46.

Takahara, S., Hamilton, H. B., Neel, J. V., Hobara, T. Y., Ogura, Y., and Nishimura, E. T. (1960). Hypocatalasemia: A new genetic state. *J. Clin. Invest.*, **39**: 610.

Talamo, R. C., Blennerhassett, J. B., and Austen, K. F. (1966). Familial emphysema and alpha₁antitrypsin deficiency. *New Engl. J. Med.*, **275**: 1301–1304.

Talamo, R. C., Langley, C. E., Reed, C. E., and Makino, S. (1973). Alpha₁-antitrypsin deficiency: A variant with no detectable alpha₁-antitrypsin. *Science*, **181**: 70–71.

Talamo, R. C., Levison, H., Lynch, M. J., Hercy, A., Hyslop, N. E., Jr., and Bain, H. W. (1971). Symptomatic pulmonary emphysema in childhood associated with hereditary alpha₁-antitrypsin and elastase inhibitor deficiency. *J. Pediatr.*, **79**: 20–26.

Tamplin, A. R., and Gofman, J. W. (1970). *Population Control through Nuclear Pollution*, Nelson-Hall Co., Chicago.

Tappel, A. L. (1965). Free radical lipid peroxidation damage and its inhibition by vitamin E and selenium. *Fed. Proc.*, **24**: 73.

Tarkoff, M. P., Kueppers, F., and Miller, W. F. (1968). Pulmonary emphysema and alpha$_1$-antitrypsin deficiency. *Am. J. Med.*, **45**: 220–228.

Tarlov, A. R., Brewer, A. J., Carson, P. E., and Alving, A. S. (1962). Primaquine sensitivity. *Arch. Int. Med.*, **109**: 209.

Tarlov, A. R., and Kellermeyer, R. W. (1961). The hemolytic effect of primaquine XI: Decreased catalase activity in primaquine-sensitivye erythrocytes. *J. Lab. Clin. Med.*, **58**: 204.

Taylor, C. V., Thomas, F. O., and Brown, M. G. (1933). Studies on protozoa: IV. Lethal effects of x-radiation on a sterile culture medium for *Culpidium campylum. Physiol. Zool.*, **6**: 467.

Taylor, D. M., Bligh, P. H., and Duggan, M. H. (1962). The absorption of calcium, strontium, barium and radium from the gastrointestinal tract of the rat. *Biochem. J.*, **83**: 25.

Teleky, L. (1943). Problems of night work: Influence on health and efficiency. *Ind. Med.*, **12**: 758.

Thiele, H. (1964). Chronic benzene poisoning. *Prac. Lek.*, **16**: 1–7.

Thier, S. O., and Segal, S. (1972). Cystinuria. In *The Metabolic Basis of Inherited Disease*, 3rd ed., J. B. Stanbury, J. B. Wyngaarden, and D. S. Fredrickson (Eds.), McGraw-Hill, New York, p. 1504.

Thoenes, W., and Bannasch, P. (1962). Elekronen- und lichtmikros-Kopische Untersuchungen am Cytoplasma der Leberzellen nach a Kuter und Chronischer Thioacetamidvergiflung. *Virchows Arch. (Pathol. Anat.)*. **335**: 556.

Thompson, D. J., Warner, S. D., and Robinson, V. B. (1974). Teratology studies on orally administered chloroform in the rat and rabbit. *Toxicol. Appl. Pharmacol.*, **28**: 348–357.

Thompson, R. P. G., Stathers, G. M., Pilcher, C. W. T., McLean, A. E. M., Robinson, J., and Williams, R. (1969). Treatment of unconjugated jaundice with dicophane. *Lancet*, **1**: 4.

Thomson, J. F. (1963). Possible role of catalase in radiation effects on mammals. *Radiat. Res.*, Suppl. **3**: 109.

Tinsley, I. J. (1969). DDT effect on rats raised on alpha-protein rations: Growth and storage of liver vitamin A. *J. Nutr.*, **98**: 319.

Tizianello, A., Pannaccivlli, I., and Salvidio, E. (1966). A simplified procedure for Brewer's methemoglobin reduction test. *Acta Haemol.*, **35**: 176.

Tobian, L., Jr. (1972). A viewpoint concerning the enigma of hypertension. *Am. J. Med.*, **52**: 595.

Toche, L., Lejeune, F., Tolat, F., Mauriquand, C., and Baron, M. M. (1960). Lead poisoning and thalassemia. *Arch. Mal. Prof.*, **21**: 329.

Toenz, O., and Betke, K. (1962). Finfacher Farbtest zur Bestimmung der Glucose-6-phosphate-dehydrogenase in menschichen Erythrocyten. *Klin. Wochenschr.*, **40**: 649.

Tokuhata, G. K. (1964). Familial factors in human lung cancer and smoking. *Am. J. Pub. Health*, **54**: 24–32.

Török, J. (1971). Pollution of drinking water and methemoglobinaemia in infants. Read

Weatherall, D. J. (1972). The thalassemias. In *The Metabolic Basis of Inherited Disease*, 3rd ed., J. B. Stanbury, J. B. Wyngaarden, and D. S. Fredrickson, McGraw-Hill, New York, p. 1432.

Weatherholtz, W. M., Campbell, T. C., and Webb, R. E. (1969). Effect of dietary protein levels on the toxicity and metabolism of heptachlor. *J. Nutr.*, **98**: 90.

Weinsier, R. L. (1976). Salt and the development of essential hypertension. *Prev. Med.*, **5**: 7–14.

Whelton, N. H., Krustev, L. P., and Billing, B. H. (1968). Reduction in serum bilirubin by phenobarbital in adult unconjugated hyperbilirubinemia: Is enzyme induction responsible? *Am. J. Med.*, **45**: 160.

White, A., and Goldstein, A. L. (1971). The role of the thymus gland in the hormonal regulation of host resistance. In *Control Process in Multicellular Organisms*, G. E. W. Wolstenholme and J. Knight (Eds.), Ciba Fdn. Symp., Churchill, London, pp. 210–237.

Whittaker, M. (1968). Differential inhibition of human serum: Cholinesterase with *n*-butyl alcohol; recognition of new phenotypes. *Acta Genet.*, **18**: 335.

Wilkenson, T. T. (1968). A review of drug toxicity in the cat. *J. Small Anim. Pract.*, **9**: 21.

Wilson, J. (1965). Skeletal abnormalities in Leber's hereditary optic atrophy. *Ann. Phys. Med.*, **8**: 91–95.

Wilson, J. (1965*a*). Leber's hereditary optic atrophy: A possible defect of cyanide metabolism. *Clin. Sci.*, **29**: 505–515.

Wilson, V. K., Thompson, M. L., and Dent, C. E. (1953). Amino-aciduria in lead poisoning. *Lancet*, **2**: 66.

Winton, E. F., and McCabe, L. J. (1970). Studies relating to water mineralization of health. *J. Am. Water Works Assoc.*, **62**(1): 26.

With, T. K. (1957). Porphyrin metabolism and barbiturate poisoning: Observations on cases of acute and chronic poisoning. *J. Clin. Path.*, **10**: 165.

With, T. K. (1963). Acute intermittent porphyria: Family studies on the excretion of PBG and delta-ALA with ion exchange chromatography. *A. Klin. Chem.*, **1**: 134.

Witkop, C. J., Jr. (1972). Albinism. In *Advances in Human Genetics*, Vol. 2, H. Harris and K. Hirschhorn (Eds.), Plenum, New York.

Witschi, H. P., and Aldridge, W. N. (1967). Biochemical changes in rat liver after acute beryllium poisoning. *Biochem. Pharmacol.*, **16**: 263–278.

Witschi, H. P. (1970). Effects of beryllium on deoxyribonucleic acid–synthesizing enzymes in regenerating rat liver. *Biochem. J.*, **120**: 623–634.

Wofsy, S. C., McElroy, M. B., and Sze, N. D. (1975). Freon consumption: Implications for atmospheric ozone. *Science*, **187**: 535–536.

Wogan, G. N. (1969). Metabolism and biochemical effects of aflatoxins. In *Aflatoxins*, Academic, New York, pp. 151–186.

Wolf, J. (1976). Softened water need not be a danger. *J. Am. Water Works Assoc.*, **68**(5): 15.

Wolf, J., and Moore, B. J. (December, 1973). Is a sodium standard necessary? *Am. Water Works Assoc. Tech. Conf. Proc.*

Wood, W. B., Jr. (1951). White blood cells v. bacteria. *Sci. Am.*, **184**: 48–52

Wynder, E. L., and Chan, P. C. (1970). The possible role of riboflavin deficiency in epithelial neoplasia: III. Induction on skin tumor development. *Cancer*, **26**: 1221–1224.

Wynder, E. L., and Chan, P. C. (1972). The possible role of riboflavin deficiency in epithelial neoplasia: III. Induction and microsomal aryl hydrocarbon hydroxylase. *J. Nat. Cancer Inst.*, **48**: 1341–1345.

Wynder, E. L., and Hoffmann, D. (1968). Experimental tobacco carcinogenesis. *Science*, **162**: 862–871.

Wynder, E. L., Hoffman, D., Chan, P., and Reddy, B. (1975). Interdisciplinary and experimental approaches: Metabolic epidemiology. In *Persons at High Risk of Cancer: An Approach to Cancer Etiology and Control*, J. F. Fraumeni, Jr. (Ed.), Academic, New York, pp. 485–500.

Wynder, E. L., Hultberg, S., Jacobsson, F., and Bross, I. J. (1957). Environmental factors in cancer of the upper alimentary tract: A Swedish study with special reference to Plummer-Vinson (Paterson-Kelly) syndrome. *Cancer*, **10**: 470–487.

Wynder, E. L., and Klin, U. E. (1965). The possible role of riboflavin deficiency in epithelial neoplasia: I. Epithelial changes in mice in simple deficiency. *Cancer*, **8**: 167–180.

Yacowitz, H., Fleischman, A. I., and Bierenbaum, M. L. (1965). Effects of oral calcium upon serum lipids in man. *Br. Med. J.*, **1**: 1352.

Yaffe, S. J., Levy, G., Matsuzawa, T., and Baliah, T. (1966). Enhancement of glucuronide-conjugating capacity in a hyperbilirubinemic infant due to apparent enzyme induction by phenobarbital. *New Engl. J. Med.*, **275**: 1461.

Yaffe, S. J., Rane, A., Sjoguist, F., Boreus, L. O., and Orrenius, S. (1970). The presence of a monooxygenese system in human fetal liver microsomes. *Life Sci.*, **9**(II): 1189–1200.

Young, T. E. (1958). Malignant melanoma in an albino. *Arch. Pathol.*, **64**: 186.

Yusipov, V. S. (1959). Effect of ascorbic acid on the carbohydrate function of the liver and the survival rate of animals with acute radiation sickness. *Meditsin. Radiol.*, **4**: 78.

Zajdela, F., and Latarjet, R. (1973). The inhibiting effect of caffeine on the induction of skin cancer by UV in the mouse. *C. R. Acad. Sci. Paris*, **29**: August.

Zamboni, L. (1965). Electron microscopic studies of blood embryogenesis in humans: I. The ultrastructure of fetal liver. *J. Ultrastruct. Res.*, **12**: 509–524.

Zannoni, V. G., and Lynch, M. M. (1973). The role of ascorbic acid in drug metabolism. *Drug Metab. Rev.*, **2**: 57–69.

Zeidberg, L. D., Prindle, R. A., and Landan, E. (1966). The Nashville air pollution study: III. Morbidity in relation to air pollution. *Am. J. Public Health*, **54**: 85–97.

Zinkham, W. H., Lenhard, R. D., and Childs, B. (1958). A deficiency of glucose-6-phosphate dehydrogenase activity in erythrocytes from patients with favism. *Bull. Johns Hopkins Hosp.*, **102**: 169.

AUTHOR INDEX

245

SUBJECT INDEX